Lecture Notes in Mathematics

An informal series of special lectures, seminars and reports on mathematical topics

Edited by A. Dold, Heidelberg and B. Eckmann, Zürich

20

Robin Hartshorne

Junior Fellow, Harvard University

Residues and Duality

Lecture Notes of a seminar on the work of A. Grothendieck, given at Harvard 1963/64

1966

Springer-Verlag · Berlin · Heidelberg · New York

Library of Congress Catalog Card Number 66-27560. Title No. 7340.

Preface

In the spring of 1963 I suggested to Grothendieck the possibility of my running a seminar at Harvard on his theory of duality for coherent sheaves — a theory which had been hinted at in his talk to the Séminaire Bourbaki in 1957 [8], and in his talk to the International Congress of Mathematicians in 1958 [9], but had never been developed systematically. He agreed, saying that he would provide an outline of the material, if I would fill in the details and write up lecture notes of the seminar. During the summer of 1963, he wrote a series of "prénotes" [10] which were to be the basis for the seminar.

I quote from the preface of the prénotes:

"Les présentes notes donnent une esquisse assez détaillée d'une théorie cohomologique de la dualité des Modules cohérents sur les préschémas. Les idées principales de la théorie m'étaient connues dès 1959, mais le manque de fondements adéquats d'Algèbre Homologique m'avait empêché d'aborder une rédaction d'ensemble. Cette lacune de fondements est sur le point d'être comblée par la thèse de VERDIER, ce qui rend en principe possible un exposé satisfaisant. Il est d'ailleurs apparu depuis qu'il existe des théories cohomologiques de dualité formellement très analogues a celle développée ici dans toutes sortes d'autres contextes: faisceaux cohérents sur les espaces analytiques, faisceaux abéliens sur les espaces topologiques (VERDIER), modules galoisiens (VERDIER, TATE), faisceaux de torsion sur les schémas munis de leur topologie étale, corps de classe en tous genres ... Cela me semble une raison assez sérieuse pour se familiariser avec le yoga général de la dualité dans un cas type, comme la théorie cohomologique des résidus.

La théorie consiste pour l'essentiel dans des questions de variance: construction d'un foncteur $f^!$ et d'un homomorphisme-trace $\underline{\underline{R}}f_*f^! \longrightarrow \mathrm{id}$. La construction donnée ici est compliquée et indirecte et n'est pas valable sous des conditions aussi générales qu'on est en droit de s'y attendre. Il faudra sans doute une idée nouvelle pour apporter des simplifications substantielles."

The seminar took place in the fall and winter of 1963-64, with the assistance of David Mumford, John Tate, Stephen Lichtenbaum, John Fogarty, and others, and gave rise to a series of six exposés which were circulated to a limited audience under the title "Séminaire Hartshorne". The present notes are a revised, expanded, and completed version of the previous notes.

I would like to take this opportunity to thank all those people who have helped in the course of this work, and in particular A. Grothendieck, who gave continual support and encouragement throughout the whole project.

R.H.

Cambridge, May 1966

CONTENTS

Introduction

ὁδὸς ανω κάτω μία καὶ ὡυτή.

— Heraclitus

The main purpose of these notes is to prove a duality theorem for cohomology of quasi-coherent sheaves, with respect to a proper morphism of locally noetherian preschemes. Various such theorems are already known. Typical is the duality theorem for a non-singular complete curve X over an algebraically closed field k, which says that

$$h^o(D) = h^1(K-D) ,$$

where D is a divisor, K is the canonical divisor, and

$$h^i(D) = \dim_k H^i(X, L(D))$$

for any i, and any divisor D. (See e.g. [16,Ch. II] for a proof.)

Various attempts were made to generalize this theorem to varieties of higher dimension, and as Zariski points out in his report [20], his generalization of a lemma of Enriques-Severi [19] is equivalent to the statement that for a normal projective variety X of dimension n over k,

$$h^{o}(D) = h^{n}(K-D)$$

for any divisor D. This is also equivalent to a theorem of Serre [FAC §76 Thm. 4] on the vanishing of the cohomology group $H^{1}(X,L(-m))$ for m large and L locally free. Using a related theorem [FAC §75 Thm. 3], Zariski shows how one can deduce on a non-singular projective variety the formula

$$h^{i}(D) = h^{n-i}(K-D)$$

for $0 \leq i \leq n$. In terms of sheaves, this result corresponds to the fact that the k-vector spaces

$$H^{i}(X,F) \quad \text{and} \quad H^{n-i}(X,F^{\vee}\otimes\omega)$$

are dual to each other, where F is a locally free sheaf, $F^{\vee}$ is the dual sheaf $\underline{Hom}(F,\mathcal{O}_X)$, and $\omega = \Omega^{n}_{X/k}$ is the sheaf of n-differentials on X. Serre [15] gives a proof of this same theorem by analytic methods for a compact complex analytic manifold X.

Grothendieck [8] gave some generalizations of these theorems for non-singular projective varieties, and then in [9] announced the general theorem for schemes proper over a field, with arbitrary singularities, which is the subject of the present lecture notes.

To motivate the statement of our main theorem, let us consider the case of projective space $X = \mathbb{P}^n_k$ over an algebraically closed field k. Then there is a canonical isomorphism

$$(1) \qquad H^n(X,\omega) \cong k$$

where $\omega = \Omega^n_{X/k}$ is the sheaf of n-differentials. Combining this with the Yoneda pairing

$$(2) \qquad H^i(X,F) \times \mathrm{Ext}^{n-i}_X(F,\omega) \longrightarrow H^n(X,\omega)$$

we obtain a pairing

$$(3) \qquad H^i(X,F) \times \mathrm{Ext}^{n-i}_X(F,\omega) \longrightarrow k$$

which one shows easily to be a perfect pairing [SGA 62, exposé 12]. This generalizes the statements above, because for a locally free sheaf F,

$$\mathrm{Ext}^{n-i}(F,\omega) = \mathrm{Ext}^{n-i}(\mathcal{O}_X, F^\vee \otimes \omega) = H^{n-i}(X, F^\vee \otimes \omega) .$$

Another way of looking at our duality pairing is as an isomorphism

$$(4) \qquad \operatorname{Ext}_X^{n-i}(F,\omega) \longrightarrow \operatorname{Hom}_k(H^i(X,F), k).$$

Since everything is linear over k, we may introduce a k-vector space G, and have an isomorphism

$$(5) \qquad \operatorname{Ext}_X^{n-i}(F, G\otimes_k\omega) \longrightarrow \operatorname{Hom}_k(H^i(X,F), G).$$

Before proceeding further, we must introduce the derived category. It will be discussed in detail in Chapter I, but for the moment it will be sufficient to know the following: For each abelian category A, there is a category D(A), called the <u>derived category</u> of A, whose objects are complexes of objects of A. If $F: A \longrightarrow B$ is an additive functor from one abelian category to another, then under reasonable conditions there is a <u>right derived functor</u> $\underline{\underline{R}}F: D(A) \longrightarrow D(B)$ with the property that for any $X \in \operatorname{Ob} A$, if X denotes also the complex which is X in degree zero, and zero elsewhere, then $H^i(\underline{\underline{R}}F(X)) = R^iF(X)$, where R^iF is the ordinary i^{th} right derived functor of F. Finally, if $F: A \longrightarrow B$ and $G: B \longrightarrow C$ are two functors then $\underline{\underline{R}}(G \cdot F) = \underline{\underline{R}}G \cdot \underline{\underline{R}}F$. This replaces the old-fashioned spectral sequence of a composite functor.

Now we can jazz up our duality for projective space as follows. We replace k by a prescheme Y, so that $X = \mathbb{P}^n_Y$. We consider the derived categories $D(X)$ and $D(Y)$ of the categories of $\mathcal{O}_X$-modules and $\mathcal{O}_Y$-modules, respectively. Then cohomology H^i becomes $\underline{\underline{R}}f_*$, the derived functor of the direct image functor f_*, where $f: X \longrightarrow Y$ is the projection. Ext becomes the derived functor $\underline{\underline{R}}$ Hom of Hom. We define $f^!(G) = f^*(G) \otimes \omega$, for $G \in D(Y)$, and we replace F by a complex of sheaves $F \in D(X)$. Then the isomorphism (1) gives us an isomorphism

$$(6) \qquad \underline{\underline{R}}f_* f^! G \xrightarrow{\sim} G$$

which we call the <u>trace map</u>. The Yoneda pairing reappears as a natural map

$$(7) \qquad \underline{\underline{R}}\, \mathrm{Hom}_X(F, f^!G) \longrightarrow \underline{\underline{R}}\, \mathrm{Hom}_Y(\underline{\underline{R}}f_*F, \underline{\underline{R}}f_*f^!G) ,$$

which, composed with the trace map (6) gives us the <u>duality morphism</u>

$$(8) \qquad \underline{\underline{R}}\, \mathrm{Hom}_X(F, f^!G) \longrightarrow \underline{\underline{R}}\, \mathrm{Hom}_Y(\underline{\underline{R}}f_*F, G)$$

which generalizes (5). This is easily proved to be an isomorphism ([III 5.1] below) under suitable hypotheses on Y,F,G. In fact, the proof is nothing but "general nonsense" once one has the isomorphism (4).

Having examined the case of projective space, we can state the following ideal theorem, which is the primum mobile of these notes, although it may never appear explicitly in this form.

Ideal Theorem. (a) For every morphism $f: X \longrightarrow Y$ of finite type of preschemes, there is a functor

$$f^{!}: D(Y) \longrightarrow D(X)$$

such that

1) if $g: Y \longrightarrow Z$ is a second morphism of finite type, then $(gf)^{!} = f^{!}g^{!}$

2) if f is a smooth morphism, then

$$f^{!}(G) = f^{*}(G) \otimes \omega ,$$

where $\omega = \Omega^{n}_{X/Y}$ is the sheaf of highest order differentials

3) if f is a finite morphism, then

$$f^{!}(G) = \underline{Hom}_{\mathcal{O}_Y}(f_{*}\mathcal{O}_X, G)^{\sim} .$$

(b) For every proper morphism $f\colon X \longrightarrow Y$ of preschemes, there is a <u>trace morphism</u>

$$Tr_f\colon \underset{=}{R}f_* f^! \longrightarrow id$$

of functors from $D(Y)$ to $D(Y)$ such that

1) if $g\colon Y \longrightarrow Z$ is a second proper morphism, then $Tr_{gf} = Tr_g Tr_f$.

2) if $X = \mathbb{P}^n_Y$, then Tr_f is the map deduced from the canonical isomorphism $R^n f_*(\omega) \cong \mathcal{O}_Y$

3) if f is a finite morphism, then Tr_f is obtained from the natural map "evaluation at one"

$$\underline{Hom}_{\mathcal{O}_Y}(f_*\mathcal{O}_X, G) \longrightarrow G .$$

c) If $f\colon X \longrightarrow Y$ is a proper morphism, then the duality morphism

$$\theta_f\colon \underset{=}{R}\, Hom_X(F, f^! G) \longrightarrow \underset{=}{R}\, Hom_Y(\underset{=}{R}f_* F, G)$$

obtained by composing the natural map (7) above with Tr_f, is an isomorphism for $F \in D(X)$ and $G \in D(Y)$.

It should be noted that we have deliberately left certain technical details out of the above statement, for ease of reading. Thus it seems reasonable to make these statements only for complexes of quasi-coherent sheaves, or rather for complexes of arbitrary sheaves, whose cohomology sheaves are quasi-coherent. (We denote this category by $D_{qc}(Y)$ or $D_{qc}(X)$.) In fact, we give an example in the case of a finite morphism to show that the duality theorem (c) fails if G is not quasi-coherent (see example following [III 6.7]). Secondly one must expect certain boundedness conditions on the complexes involved, namely F should be bounded above (we write $F \in D^-_{qc}(X)$) and G should be bounded below ($G \in D^+_{qc}(Y)$) for the duality theorem. Finally, questions of variance will be very important in the proof, and so the equals signs in a1 and b1 must be taken with a grain of salt. We must preserve very carefully the distinction between "equals" and "is canonically isomorphic to". Hence we will have more precise statements below.

As to proving the ideal theorem, we succeed only partially. Certainly the conditions under which we can prove it will suffice for most applications, but it is unsatisfying to have restrictive hypotheses which are apparently not essential to the truth of the theorem. I mention four sets of hypotheses under which the theorem can be proved.

(i) For the category of noetherian preschemes of finite Krull dimension, and morphisms $f: X \longrightarrow Y$ which can be factored through a suitable projective space $\mathbb{P}^N_Y$, we have a1-3 for $D^+_{qc}(Y)$ [III 8.7], b1-3 for $D^+_{qc}(Y)$ [III 10.5], and c for $F \in D^-_{qc}(X)$ and $G \in D^+_{qc}(Y)$ [III 11.1].

(ii) For the category of noetherian preschemes which admit dualizing complexes (see [V §10]; this implies in particular that the preschemes have finite Krull dimension) and morphisms whose fibres are of bounded dimension, we have a1-3 and b1-3 for $D^+_c(Y)$, and c for $F \in D^-_{qc}(X)$, $G \in D^+_c(Y)$ [VII 3.4]. Here the subscript "c" denotes complexes with <u>coherent</u> cohomology sheaves.

(iii) For the category of noetherian preschemes of finite Krull dimension and <u>smooth</u> morphisms, we have a1-3 for $D^+_{qc}(Y)$, b1-3 for $G \in D^b_{qc}(Y)$, and c for $F \in D^-_{qc}(X)$ and $G \in D^b_{qc}(Y)$ [VII 4.3]. Here the exponent "b" denotes complexes which are bounded in both directions, i.e., finite.

(iv) Recently P. Deligne has shown (unpublished) that for the category of noetherian preschemes, one can construct $f^!$ and Tr_f satisfying a1,b1, and c, working with $F \in D(\mathrm{Qco}(X))$ and $G \in D^+(\mathrm{Qco}(Y))$, the derived categories of the categories of quasi-coherent sheaves on X and Y, respectively.

The principal difficulty has been the lack of a suitable construction of the functor $f^!$. Our procedure is to define it locally, and then glue. For a finite or a smooth morphism, we have it given to us, by a2 and a3 [III §2,§6]. Thus by composition we can obtain it for any morphism which can be factored into a finite morphism followed by a smooth morphism [III §8]. However, the derived category is not a local object, so we cannot glue these local determinations of $f^!$ to obtain a global one. We resort to a clumsy, round-about procedure of defining $f^!$ for a special class of complexes, called residual complexes [VI §3], and then pulling ourselves up by our bootstraps to get it for arbitrary complexes [VII §3]. But our result has the unpleasant hypotheses of (ii) above.

Deligne's construction of $f^!$ is entirely different, and proceeds essentially by representing the functor (for $F \in D(X)$)

$$F \longmapsto \underline{\underline{R}}\,\mathrm{Hom}_Y\,(\underline{\underline{R}}f_*F, G).$$

This approach has the advantage of giving the duality theorem immediately. He also has a method for calculating $f^{!}G$ locally on X. However, it is not immediately clear from his construction that the properties a2,3 and b2,3 hold. In fact, their proof will probably require some knowledge of the duality theorem as we have proved it here. At least one can hope that some combination of the two approaches will provide substantial simplifications of the theory at a later date.

A second difficulty, which lurks on the fringes of these notes, is that the derived category seems to be a little too big when it comes to unbounded complexes. For example, one can have two unequal morphisms $f,g: X \longrightarrow Y$ in $D^{+}(A)$ (where A is an abelian category) such that their restrictions to each truncation of the complex X to a bounded complex, are equal. This gives rise to some trouble with unbounded complexes (see the problem after [II 5.7], the boundedness hypotheses in [IV 3.1] and [VII 4.3a], and the remark following [VI 1.1c]). Perhaps one will have to replace D^{+} by the categories ind D^{b} and pro D^{b}, which however may not be triangulated categories!

Thirdly, some discussion of our noetherian hypotheses is in order. In the present state of the theory, the noetherian hypotheses are well entrenched. We have used them in [II §7] for the structure of the injective objects in the category of sheaves on a locally noetherian prescheme, and its consequence that $D^+(Qco(X)) \longrightarrow D^+_{qc}(X)$ is an equivalence of categories. We have used them in the construction of the trace map for projective space [III §4], in the construction of $f^!$ for a finite morphism [III §6], in the whole theory of dualizing complexes [V §2], in the definition of residual complexes [VI §1], and so forth. Our finite Krull dimension hypotheses are often needed only to make possible the definition of $\underline{\underline{R}}f_*$ for unbounded complexes — a problem which will disappear when the second difficulty above is solved, and the relation between $D(Qco(X))$ and $D_{qc}(X)$ is better understood. It is certainly clear that our methods of proof rely heavily on noetherian hypotheses. I expect, however, that once a suitable statement is obtained, e.g., in case (iii) above, one could expect to prove the theorem without noetherian hypotheses, by reducing to the noetherian case. More satisfactory, of course, would be a treatment where noetherian hypotheses on the base were eliminated from the proofs as well as from the statements. When that is achieved, it will be reasonable to state the duality theorem for a proper morphism of ringed topos, whose fibres are noetherian schemes... At the present, however, this must remain a dream of things to come.

Now we will give the reader a brief description of the organization of these notes.

Chapter I gives the language of derived categories, which is used continually in the sequel. Sections 1-5, containing the definition of derived categories and derived functors, are essential, while sections 6 and 7 are refinements which are needed in proofs later on. This chapter is a self-contained treatment of the subject, and makes no reference to algebraic geometry, so can be used independently as an introduction to the notion of derived category. Sections 1-6 (except for the notion of localizing subcategory and the corresponding categories $K_{A'}(A)$, $D_{A'}(A)$, etc.) are taken almost without change from notes of Verdier, and should appear in his thesis [18].

Chapter II is a fairly systematic treatment of the applications of the language of derived categories to the category of sheaves on a prescheme. We consider the functors Γ , f_* , Hom , Hom , $\otimes$, f^* , their derived functors, and relations between these derived functors, such as associativity formulae. There is really no new mathematics involved here, since it is merely a translation into a new language of known results. However, some care is taken to see what hypotheses are needed. Only section 7 is notable new material. Here we give the structure of injective $\mathcal{O}_X$-modules on a locally noetherian prescheme X, showing that they are all direct sums of certain indecomposable injectives $J(x,x')$, for pairs of points x specializing to x' of X. This extends results of Matlis [13] and Gabriel [5] for the case of quasi-coherent sheaves.

Chapter III contains everything we can say about duality for projective morphisms, and if the reader cares only for them, he may stop at the end of this chapter. However, it should be noted that the following chapters, besides giving us the tools for the proof of the general duality theorem in Chapter VII, also give much new insight into the nature of duality for a projective morphism.

There are so many situations in which we have a functor $f^{!}$ like the one mentioned in the ideal theorem above, that we use different notations for them. Thus we have $f^{\#}$ for a smooth morphism in section 2, $f^{\flat}$ for a finite morphism in section 6, and $f^{!}$ for an embeddable morphism in section 8. Later we will also have f^{Y}, f^{Z} [VI §2] and f^{Δ}[VI §3] for residual complexes.

In sections 3, 4, and 5 we recall the explicit calculations of cohomology for projective space, and give the old duality for projective space in the new language of derived categories. Sections 6 and 7 give the corresponding formalism for finite morphisms, and its relation to the case of smooth morphisms, so that in sections 8, 10, and 11 we can prove the ideal duality theorem for projective morphisms. Section 9 gives the formalism of a residue symbol which generalizes the classical residue of a differential on a curve. Even over the complex numbers, this important concept of residue for varieties of dimension greater than one was not known before.

In Chapter IV we study "local cohomology", or cohomology with restricted supports, of abelian sheaves on a locally noetherian topological space, generalizing results of [LC §§1,3]. In particular, we discuss various cohomological properties which a sheaf or complex of sheaves may have with respect to certain families of supports. This gives rise to the notions of depth, Cousin complex, Cohen-Macaulay complex, and Gorenstein complex. The results of this chapter are independent of all other chapters of these notes, and so may be of use elsewhere, although their only application for the moment is to the theory of duality on preschemes.

Chapter V discusses dualizing complexes (read section 0 to find out what one is) and gives a duality theorem for modules over a local ring, generalizing results of [LC §§2,4,6]. In particular, the dualizing functor $\underline{D} = \underline{\underline{R}}\ \underline{Hom}(\cdot,R^{\cdot})$ treated here will be useful for our bootstrap operation (construction of $f^{!}$) later, because it interchanges $\otimes$ and $\underline{Hom}$, $f^{!}$ and $\underline{\underline{L}}f_{*}$, and commutes with $\underline{\underline{R}}f_{*}$ (duality theorem!).

In Chapter VI we prepare for the final duality theorem by giving the construction of $f^{!}$ and Tr_f for residual complexes. This is accomplished by a delicate glueing procedure which is the most difficult part of the theory, so we have given it in some detail. Perhaps some day this type of construction will be done more elegantly using the language of fibred categories and results of Giraud's thesis [6].

Chapter VII contains two main results. The first is the residue theorem, which generalizes the classical theorem that the sum of the residues of a differential on a curve is zero, and which is proved by reduction to that case. The second is the proof of the duality theorem for a proper morphism, which, now that all the functorial machinery has been set up, is little more than putting together the pieces of a jigsaw puzzle.

It remains to give the reader some perspective by listing some topics which have a logical place in these notes, but which are <u>not</u> here.

1. The cohomology class associated to a cycle. Let X be proper and smooth of dimension n over a field k. In [8, §4] it is shown how to associate to each non-singular subvariety Y of codimension p of X, a cohomology class

$$P_X(Y) \in H^p(X, \Omega^p_{X/k}) .$$

Now this can be done for an arbitrary subvariety of X, using the remarks in [9]. One defines $\omega_Y = H^{-n+p}(g^! k)$, where $g\colon Y \longrightarrow k$ is the projection. Then there is a canonical element

$$\eta \in \operatorname{Ext}^{-n+p}_{\mathcal{O}_Y}(\omega_Y, g^! k) .$$

One defines a natural map

$$\Omega^{n-p}_{Y/k} \longrightarrow \omega_Y ,$$

which, together with η and the construction of [8, §4], gives the cohomology class $P_X(Y)$. One proves the fundamental theorem that formation of the cohomology class associated to a cycle takes an intersection of cycles into the cup-product of their cohomology classes.

2. The theory of Poincaré duality and the Gysin homomorphism can be developed as in [8, §7].

3. A Lefschetz-Verdier fixed point formula for coherent sheaves on a scheme proper over a field, to generalize [21, Thm. 2]. In particular, the determination of the local contribution at a non-simple fixed point presents an interesting topic for future investigation.

4. The still lacking theory of duality for complexes with differential operators as boundary operator, its ties with ordinary singular <u>homology</u> theory and with vector bundles with integrable connections, as suggested in [22].

A remark on references: Theorem 5.1 of Chapter III is referred to as "Theorem 5.1" in Chapter III, and as [III 5.1] elsewhere. References to the bibliography at the end are given by square brackets with an arabic numeral or some capital letters e.g., [14,(31.1)] or [EGA III 2.1.12].

CHAPTER I. THE DERIVED CATEGORY

§0. Introduction.

Let A and B be abelian categories, and let $F: A \longrightarrow B$ be a functor. Our purpose is to define, functorially for each $X \in \mathrm{Ob}\, A$, a complex $\underline{\underline{R}}F(X)$, whose cohomology groups are the right derived functors of F acting on X, $R^i F(X)$. In general, we will define $\underline{\underline{R}}F(X^{\cdot})$ for any complex $X^{\cdot}$ of objects of A.

More precisely, $\underline{\underline{R}}F$ will be a functor from the derived category D(A) of A, to the derived category D(B) of B. The derived category D(A) is obtained as follows: one first considers the category K(A), whose objects are complexes of elements of A, and whose morphisms are homotopy equivalence classes of morphisms of complexes. The category D(A) is obtained by "localizing" K(A), so that every morphism in K(A) which induces an isomorphism on cohomology, becomes an isomorphism in D(A). This process of localization will be explained in general below.

Although A and B above are assumed to be abelian categories, the categories K(A), D(A)... are in general not abelian. However, they can be given a structure which carries enough information for our purposes, namely a structure of triangulated category. Thus we will be led to study triangulated categories.

§1. Triangulated Categories.

Definition. A *triangulated category* is an additive category C, together with

a) an automorphism $T: C \longrightarrow C$ of the category called the *translation functor*, and

b) a collection of sextuples (X,Y,Z,u,v,w), called the *triangles of* C, where in each sextuple, X,Y,Z are objects of C, and u,v,w are morphisms as follows: $u: X \longrightarrow Y$, $v: Y \longrightarrow Z$, $w: Z \longrightarrow T(X)$. A triangle is usually written

$$\begin{array}{ccc} & Z & \\ {}^{w}\swarrow & & \nwarrow^{v} \\ X & \xrightarrow{\ u\ } & Y \end{array} .$$

A *morphism* of triangles $(X,Y,Z,u,v,w) \longrightarrow (X',Y',Z',u',v',w')$ is a commutative diagram

$$\begin{array}{ccccccc} X & \xrightarrow{u} & Y & \xrightarrow{v} & Z & \xrightarrow{w} & T(X) \\ \downarrow f & & \downarrow g & & \downarrow h & & \downarrow T(f) \\ X' & \xrightarrow{u'} & Y' & \xrightarrow{v'} & Z' & \xrightarrow{w'} & T(X') \end{array} .$$

This data is subject to the following axioms:

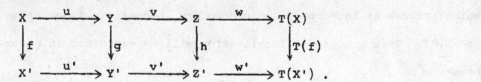

(TR1) Every sextuple (X,Y,Z,u,v,w) as above, isomorphic to a triangle, is a triangle. Every morphism $u: X \longrightarrow Y$ can be imbedded in a triangle (X,Y,Z,u,v,w). The sextuple $(X,X,0,id_X,0,0)$ is a triangle.

(TR2) (X,Y,Z,u,v,w) is a triangle if and only if $(Y,Z,T(X),v,w,-T(u))$ is.

(TR3) Given two triangles (X,Y,Z,u,v,w) and (X',Y',Z',u',v',w'), and morphisms $f: X \longrightarrow X'$, $g: Y \longrightarrow Y'$ commuting with u,u', there exists a morphism $h: Z \longrightarrow Z'$ (not necessarily unique!) so that (f,g,h) is a morphism of the first triangle into the second.

(TR4) (The octohedral axiom).

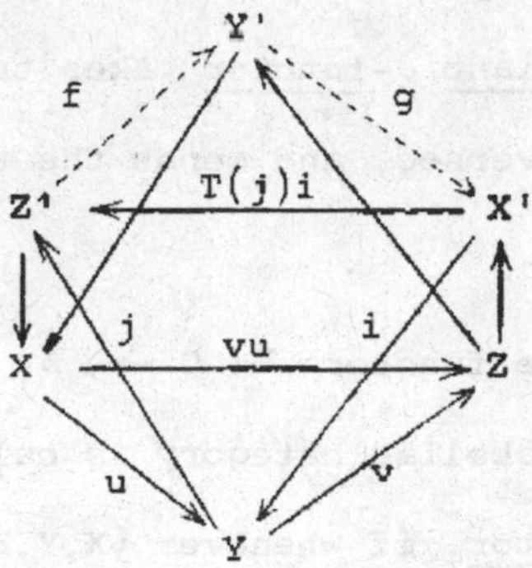

Suppose given triangles

$$(X, Y, Z', u, j, .)$$
$$(Y, Z, X', v, ., i)$$
$$(X, Z, Y', vu, ., .) \quad .$$

Then there exist morphisms $f: Z' \longrightarrow Y'$ and $g: Y' \longrightarrow X'$, such that

$$(Z', Y', X', f, g, T(j)i)$$

is a triangle, and the two other faces of the octohedron with f,g as edges, are commutative diagrams.

Definition. An additive functor $F: C \rightarrow C'$ from one triangulated category to another is called a (covariant) ∂-functor if it commutes with the translation functor and takes triangles into triangles. A contravariant ∂-functor takes triangles into triangles with the arrows reversed, and sends the translation functor into its inverse.

Definition. An additive functor $H: C \longrightarrow A$ from a triangulated category to an abelian category is called a covariant cohomological functor, if whenever (X,Y,Z,u,v,w) is a triangle, the long sequence

$$\cdots \longrightarrow H(T^iX) \longrightarrow H(T^iY) \longrightarrow H(T^iZ) \longrightarrow H(T^{i+1}X) \longrightarrow \cdots$$

is exact (the morphisms being $H(T^iu)$ etc.). If H is a cohomological functor, we often write $H^i(X)$ for $H(T^iX)$, $i \in \mathbb{Z}$. One defines a <u>contravariant cohomological functor</u> by reversing the arrows.

<u>Proposition 1.1.</u> a) The composition of any two consecutive morphisms in a triangle is zero.

b) If C is a triangulated category, and M an object of C, then $\mathrm{Hom}_C(M,\cdot)$ and $\mathrm{Hom}_C(\cdot,M)$ are cohomological functors on C.

c) If in the situation of (TR3) f and g are isomorphisms, then h is also an isomorphism.

<u>Proof</u>. a) Let (X,Y,Z,u,v,w) be a triangle. By (TR2) it is sufficient to show that $vu = 0$. Also by (TR2), $(Y,Z,T(X),v,w,-T(u))$ is a triangle. By (TR1), $(Z,Z,0,\mathrm{id}_Z,0,0)$ is a triangle. We apply (TR3) to the maps $v\colon Y \longrightarrow Z$ and $\mathrm{id}_Z\colon Z \longrightarrow Z$, and conclude that there is a map $h\colon T(X) \longrightarrow 0$ giving a morphism of triangles. It follows that $T(v)(-T(u)) = 0$, or, since T is an automorphism, $vu = 0$.

b) Let $M \in \mathrm{Ob}\, C$, and let (X,Y,Z,u,v,w) be a triangle. To show $\mathrm{Hom}_C(M,\cdot)$ is a cohomological functor, it will be sufficient by (TR2) to show the sequence

$$\mathrm{Hom}_C(M,X) \longrightarrow \mathrm{Hom}_C(M,Y) \longrightarrow \mathrm{Hom}_C(M,Z)$$

is exact. By a) we know the composition is zero. So suppose given $g \in \mathrm{Hom}_C(M,Y)$ such that $vg \in \mathrm{Hom}_C(M,Z)$ is zero. We apply (TR3) to the triangles $(M,M,0,\mathrm{id}_M,0,0)$ and (X,Y,Z,u,v,w) and the map $g\colon M \to Y$ and $0\colon 0 \to Z$ and conclude that there exists an $f\colon M \to X$ such that $uf = g$.

A similar proof shows that $\mathrm{Hom}_C(\cdot,M)$ is a (contravariant) cohomological functor.

c) In the situation of (TR3) suppose that f and g are isomorphisms. We apply the cohomological functor $\mathrm{Hom}_C(Z',\cdot)$ to the whole situation, and obtain an exact commutative diagram

$$\begin{array}{ccccccccc}
\mathrm{Hom}(Z',X) & \to & \mathrm{Hom}(Z',Y) & \to & \mathrm{Hom}(Z',Z) & \to & \mathrm{Hom}(Z',T(X)) & \to & \mathrm{Hom}(Z',T(Y)) \\
\downarrow f_* & & \downarrow g_* & & \downarrow h_* & & \downarrow T(f)_* & & \downarrow T(g)_* \\
\mathrm{Hom}(Z',X') & \to & \mathrm{Hom}(Z',Y') & \to & \mathrm{Hom}(Z',Z') & \to & \mathrm{Hom}(Z',T(X')) & \to & \mathrm{Hom}(Z',T(Y'))
\end{array}$$

where $f_* = \mathrm{Hom}(Z',f)$ etc. Now since f and g are isomorphisms in C, it follows that $f_*, g_*, T(f)_*$, and $T(g)_*$ are isomorphisms of abelian groups. Hence by the five-lemma, h_* is an isomorphism. We conclude that there exists a $\varphi \in \mathrm{Hom}_C(Z',Z)$ such that $h_*(\varphi) = h\varphi$ is $\mathrm{id}_{Z'} \in \mathrm{Hom}(Z',Z')$.

Similarly using the cohomological functor $\mathrm{Hom}_C(\cdot,Z)$ we find there is a $\psi \in \mathrm{Hom}(Z',Z)$ such that $\psi h = \mathrm{id}_{Z}$. It follows that $\varphi = \psi$ and h is an isomorphism.

§2. K(A) is triangulated.

Let A be an abelian category. A complex of objects of A is a collection $X^{\cdot} = (X^n)_{n\in\mathbb{Z}}$ of objects of A, together with maps $d^n\colon X^n \longrightarrow X^{n+1}$ such that $d^{n+1}d^n = 0$ for all $n \in \mathbb{Z}$. A morphism f of complexes $X^{\cdot}$ to $Y^{\cdot}$ is a collection of maps $f^n\colon X^n \longrightarrow Y^n$ which commute with the maps of complexes:

$$f^{n+1}d_X^n = d_Y^n f^n$$

for all n. Two maps $f,g\colon X^{\cdot} \longrightarrow Y^{\cdot}$ are said to be homotopic if there is a collection of maps $k = (k^n)$, $k^n\colon X^n \longrightarrow Y^{n-1}$ (which do not necessarily commute with d) such that

$$f^n - g^n = d_Y^{n-1}k^n + k^{n+1}d_X^n$$

for all n. Homotopy is an equivalence relation, and the compositions of homotopic maps are homotopic.

We define K(A) to be the category whose objects are complexes of objects of A, and whose morphisms are homotopy equivalence classes of morphisms of complexes. A complex $X^{\cdot}$ is said to be bounded below if $X^n = 0$ for $n \ll 0$. We denote by $K^+(A)$ the full subcategory of K(A) consisting of the complexes bounded below. Similarly we define $K^-(A)$ and $K^b(A)$ by taking complexes bounded above, or bounded on both sides, respectively.

Now let us define the structure of triangulated category on $K(A)$ (resp. $K^+(A)$, etc.). T will be the operation of shifting one place to the left, and changing the sign of the differential, i.e., $T(X^\cdot)^p = X^{p+1}$, and $d_{T(X)} = -d_X$. We will often write $X^\cdot[1]$ instead of $T(X^\cdot)$, and $X^\cdot[n]$ instead of $T^n(X^\cdot)$. If $u: X^\cdot \longrightarrow Y^\cdot$ is any morphism, consider the mapping cone $Z^\cdot$ of u. Recall that the mapping cone is defined to be the complex $T(X^\cdot)\oplus Y^\cdot$, where the differential operator is given by the matrix

$$\begin{pmatrix} T(d_X) & T(u) \\ 0 & d_Y \end{pmatrix}.$$

There are natural morphisms $v: Y^\cdot \longrightarrow Z^\cdot$ and $w: Z^\cdot \longrightarrow T(X^\cdot)$.

We define a triangle in $K(A)$ to be any sextuple isomorphic to a sextuple from a morphism $u: X^\cdot \longrightarrow Y^\cdot$ by the construction above.

All of the axioms (TR1)-(TR4) are easy to verify, once one has made the observation that the mapping cone of the identity map $id_X: X^\cdot \longrightarrow X^\cdot$ is homotopic to 0. Indeed, the mapping cone is $T(X^\cdot)\oplus X^\cdot$ as described above. The matrix

$$k = \begin{pmatrix} 0 & 0 \\ id_X & 0 \end{pmatrix}$$

is a homotopy operator.

<u>Definition</u>. We define $\underline{H}$ to be the functor from $K(A)$ to A which takes a complex $X^{\cdot}$ into its 0th cohomology group, namely $\ker d^0/\operatorname{im} d^{-1}$. (This is indeed a functor, because homotopic maps of complexes induce the same map on cohomology.) We write H^i for $H \cdot T^i$, for any $i \in \mathbb{Z}$.

Observe that H is a cohomological functor from $K(A)$ to A. Indeed, it is sufficient to check the long exact sequence for triangles constructed with the mapping cylinder of a morphism $u: X^{\cdot} \longrightarrow Y^{\cdot}$ of complexes, and there one can check directly that the sequence is exact.

§3. Localization of Categories.

Definition. Let C be a category. A collection S of arrows of C is called a multiplicative system if it satisfies the following axioms (FR1)-(FR3):

(FR1) If $f,g \in S$, and fg exists, then $fg \in S$. For any $X \in \operatorname{Ob} C$, $id_X \in S$.

(FR2) Any diagram

$$\begin{array}{ccc} & & Z \\ & & \big\downarrow{\scriptstyle s} \\ X & \xrightarrow{u} & Y \end{array}$$

with $s \in S$ can be completed to a commutative diagram

$$\begin{array}{ccc} W & \xrightarrow{v} & Z \\ {\scriptstyle t}\big\downarrow & & \big\downarrow{\scriptstyle s} \\ X & \xrightarrow{u} & Y \end{array}$$

with $t \in S$. Ditto for the opposed statement (i.e., with all arrows reversed).

(FR3) If $f,g: X \longrightarrow Y$ are morphisms in C, the following conditions are equivalent:

(i) There exists an $s: Y \longrightarrow Y'$ in S such that $sf = sg$.

(ii) There exists a $t: X' \longrightarrow X$ in S such that $ft = gt$.

<u>Definition</u>. If C is a category, and S a collection of morphisms of C, then the <u>localication of C with respect to S</u> is a category C_S, together with a functor $Q: C \longrightarrow C_S$ such that

a) $Q(s)$ is an isomorphism for every $s \in S$, and

b) Any functor $F: C \longrightarrow D$ such that $F(s)$ is an isomorphism for all $s \in S$ factors uniquely through Q.

<u>Remark</u>. One can show that such a localization exists without hypotheses on S, but we will not need this result.

<u>Proposition 3.1</u>. Let C be a category, and S a multiplicative system in C. Then we can obtain the localization C_S as follows: $\mathrm{Ob}\, C_S = \mathrm{Ob}\, C$, and for any $X,Y \in \mathrm{Ob}\, C$,

$$\mathrm{Hom}_{C_S}(X,Y) = \varinjlim_{I_X} \mathrm{Hom}_C(X',Y)$$

where I_X is the category whose objects are morphisms $s: X' \longrightarrow X$ in S, and whose morphisms are commutative diagrams

$$\begin{array}{ccc} X' & \xrightarrow{\ f\ } & X'' \\ & {\scriptstyle s_1}\searrow \quad \swarrow{\scriptstyle s_2} & \\ & X & \end{array}$$

.

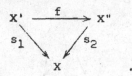

Furthermore, if C is an additive category, so is C_S.

Proof. First observe, using (FR1), (FR2), and (FR3) that the category I_X satisfies the axioms L1, L2, L3 of [GT, Ch. I, §1], and hence behaves as well as an inductive system for taking limits. Thus a morphism of X to Y in C_S is represented by a diagram

$$X \xleftarrow{\ s\ } X' \xrightarrow{\ a\ } Y$$

with $s \in S$. This diagram defines the same morphism as another one with $t \in S$

$$X \xleftarrow{\ t\ } X'' \xrightarrow{\ b\ } Y$$

if and only if there is a morphism $u: X''' \longrightarrow X$ in S and morphisms $f: X''' \longrightarrow X'$ and $g: X''' \longrightarrow X''$ such that $sf = u = tg$ and $af = bg$.

To compose morphisms

$$X \xleftarrow{\ s\ } X' \xrightarrow{\ a\ } Y \qquad \text{and} \qquad Y \xleftarrow{\ t\ } Y' \xrightarrow{\ b\ } Z$$

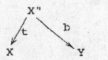

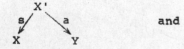

we use (FR2) to find a commutative diagram

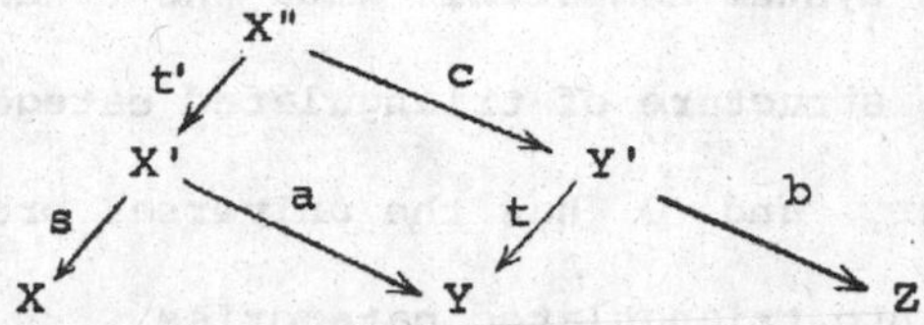

with $t' \in S$, and then take $X"$, st', bc to be the composition. One verifies easily that the resulting morphism of X to Z is independent of the representatives of the morphisms of X to Y and Y to Z chosen, and is also independent of the commutative diagram chosen.

One can also verify easily that the functor $Q: C \longrightarrow C_S$ has the properties required and that C_S is additive if C is (again using L1, L2, L3 to show that the $\varinjlim$ is a group).

Definition. Let C be a triangulated category and S a multiplicative system of morphisms. S is said to be compatible with the triangulation if the following two axioms are satisfied:

(FR4) $s \in S \Longleftrightarrow T(s) \in S$, where T is the translation functor.

(FR5) The same as (TR3), but where we assume that $f, g \in S$, and require that $h \in S$.

<u>Proposition 3.2</u>. If C is a triangulated category and S is a multiplicative system compatible with the triangulation, then C_S has a unique structure of triangulated category such that Q is a ∂-functor, and Q has the universal property b) above for ∂-functors into triangulated categories.

<u>Proof</u>. Left to the reader. It helps to observe that one can also calculate $\mathrm{Hom}_{C_S}(X,Y)$ as

$$\varinjlim_{J_Y} \mathrm{Hom}_C(X,Y')$$

where J_Y is the category whose objects are morphisms $s\colon Y \longrightarrow Y'$ in S, and whose maps are commutative diagrams

$$\begin{array}{ccc} & Y & \\ {}^{s_1}\swarrow & & \searrow^{s_2} \\ Y' & \xrightarrow{\ f\ } & Y'' \end{array} .$$

<u>Proposition 3.3</u>. Let C be a category, and let S be a multiplicative system in C. Let D be a full subcategory of C (i.e., $X,Y \in \mathrm{Ob}\, D \Longrightarrow \mathrm{Hom}_D(X,Y) = \mathrm{Hom}_C(X,Y)$) and assume that $S \cap D$ is a multiplicative system in D. Assume furthermore that one of the following two conditions holds:

(i) Whenever $s: X' \longrightarrow X$ is a morphism in S, with $X \in \mathrm{Ob}\, D$, then there is a morphism $f: X'' \longrightarrow X'$ such that $X'' \in \mathrm{Ob}\, D$ and $sf \in S$.

(ii) Ditto with the arrows reversed.

Then the natural functor $D_{S \cap D} \longrightarrow C_S$ is fully faithful, i.e., $D_{S \cap D}$ can be identified with a full subcategory of C_S.

<u>Proof</u>. Straightforward.

<u>Proposition 3.4</u>. Let C be a category, S a multiplicative system in C, and $Q: C \longrightarrow C_S$ the localization functor. Let D be another category, and let $F, G: C_S \longrightarrow D$ be two functors. Then the natural map

$$\alpha: \ \mathrm{Hom}(F,G) \longrightarrow \mathrm{Hom}(FQ,GQ)$$

of morphisms of functors is bijective.

<u>Proof</u>. To give a morphism of functors $F \longrightarrow G$ means to give, for each $X \in \mathrm{Ob}\, C_S$, a morphism $F(X) \longrightarrow G(X)$, such that if $X \longrightarrow Y$ is a morphism, then

$$\begin{array}{ccc} F(X) & \longrightarrow & F(Y) \\ \downarrow & & \downarrow \\ G(X) & \longrightarrow & G(Y) \end{array}$$

is a commutative diagram. Thus, since $\text{Ob } C = \text{Ob } C_S$, the map α is injective. To show α surjective, suppose given a morphism $FQ \longrightarrow GQ$. Then we have a morphism $F(X) \longrightarrow G(X)$ for each $X \in \text{Ob } C = \text{Ob } C_S$, and commutative diagrams for morphisms $X \longrightarrow Y$ in C. A morphism in C_S is represented by a diagram

$$\begin{array}{ccc} & X' & \\ {}^{s}\swarrow & & \searrow{}^{f} \\ X & & Y \end{array}$$

of morphisms in C, with $s \in S$. But for $s \in S$, $F(s)$ and $G(s)$ are isomorphisms, so we get the required commutative diagram.

§4. Qis and the Derived Category.

Let A be an abelian category, and let $K(A)$ be the triangulated category described in §2. We define a *quasi-isomorphism* to be a morphism $f: X^{\cdot} \longrightarrow Y^{\cdot}$ in $K(A)$ which induces an isomorphism on cohomology. Let *Qis* be the collection of all quasi-isomorphisms.

Proposition 4.1. Qis is a multiplicative system in $K(A)$.

Proof. This is a consequence of the following more general proposition.

Proposition 4.2. Let C be a triangulated category, let A be an abelian category, and let H be a cohomological functor from C to A. Let S be the set of morphisms s in C such that $H(T^i(s))$ is an isomorphism for all $i \in \mathbb{Z}$. Then S is a multiplicative system in C, compatible with the triangulation.

Proof. We must verify the axioms (FR1)-(FR5). (FR1) and (FR4) are trivial. (FR5) follows from the long exact sequence of a cohomological functor and the five-lemma.

To prove (FR2), let a diagram

$$\begin{array}{ccc} & & Z \\ & & \downarrow{\scriptstyle s} \\ X & \xrightarrow{\;u\;} & Y \end{array}$$

be given, with $s \in S$. Complete s to a triangle (Z,Y,N,s,f,g). Complete fu to a triangle (W,X,N,t,fu,h). Then (u, id_N) is a map of two sides of the second triangle into the first, so there is a map $v: W \longrightarrow Z$ giving a morphism of triangles.

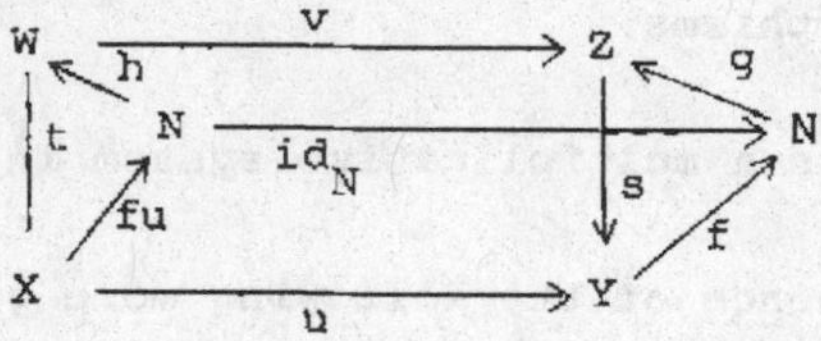

.

Now $sv = ut$, so it remains to prove $t \in S$. Indeed, since $s \in S$, we have $H(T^i(N)) = 0$ for all $i \in \mathbb{Z}$ by the long exact sequence of the first triangle. Applying this to the long exact sequence of the second triangle, we find $H(T^i(t))$ is an isomorphism for all $i \in \mathbb{Z}$.

The opposed statement of (FR2) is proved similarly.

To prove (FR3), we consider the morphism $f-g$, and reduce to showing the following two properties equivalent (where $f: X \longrightarrow Y$ is a morphism):

(i) There exists an $s: Y \longrightarrow Y' \in S$ such that $sf = 0$

(ii) There exists a $t: X' \longrightarrow X \in S$ such that $ft = 0$.

Suppose (i) holds.

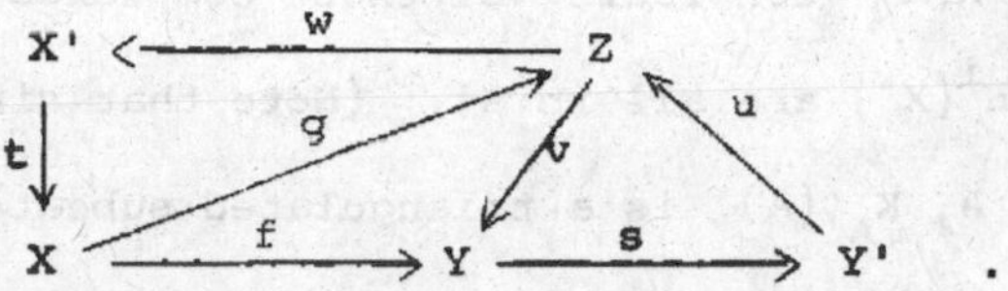

By (TR1) and (TR2) we can find a triangle (Z,Y,Y',v,s,u) for suitable Z. Now $sf = 0$, so by Proposition 1.1 b), there is a map $g: X \longrightarrow Z$ such that $f = vg$. Again by (TR1) and (TR2) we can find a triangle (X',X,Z,t,g,w) for suitable X'. By the same Proposition applied to this second triangle, the existence of v implies that $ft = 0$. We need only show that $t \in S$. Since $s \in S$, $H(T^i(Z)) = 0$ for all $i \in \mathbb{Z}$, by the long exact sequence of cohomology. In turn, this implies that $t \in S$.

The implication (ii) $\Longrightarrow$ (i) is analogous.

<u>Definition</u>. The <u>derived category</u> of A, D(A), is defined to be $K(A)_{Qis}$. Similarly we define $D^+(A) = K^+(A)_{Qis}$, $D^-(A)$, and $D^b(A)$. One checks easily using Proposition 3.3 that they are full subcategories of D(A), and that $D^+(A) \cap D^-(A) = D^b(A)$.

<u>Definition</u>. Let A be an abelian category, and let A' be a <u>thick</u> abelian subcategory of A (i.e., any extension in A of two objects of A' is in A'). We define $K_{A'}(A)$ to be the full subcategory of $K(A)$ consisting of those complexes $X^{\cdot}$ whose cohomology objects $H^i(X^{\cdot})$ are all in A'. (Note that since A' is a thick subcategory of A, $K_{A'}(A)$ is a triangulated subcategory of $K(A)$, i.e., if two sides of a triangle are in it, so is the third.) We define $D_{A'}(A)$ to be $K_{A'}(A)_{Qis}$. Note by Proposition 3.3 that $D_{A'}(A)$ is the full subcategory of $D(A)$ consisting of those $X^{\cdot}$ with all $H^i(X^{\cdot}) \in A'$. We define similarly $K^+_{A'}(A)$, $D^+_{A'}(A)$ by taking complexes bounded below, with cohomology in A', etc.

<u>Remark</u>. There is a natural functor $D(A') \longrightarrow D_{A'}(A)$ which in general is neither injective nor surjective. (See however Proposition 4.8.)

<u>Example</u>. To help understand the category $D(A)$, let us ask the question, when does a morphism of complexes $f: X \longrightarrow Y$ give the zero map in $D(A)$? The condition is the following:

(*) There exists an $s: Y \longrightarrow Y'$ in Qis such that sf is homotopic to zero (or, equivalently, there exists a $t: X' \longrightarrow X$ in Qis such that ft is homotopic to zero).

1. Of course, if $f \sim 0$ (f homotopic to zero), then f satisfies (*). The converse is false. For example, take X to be the complex

$$0 \longrightarrow \mathbb{Z} \xrightarrow{2} \mathbb{Z} \longrightarrow \mathbb{Z}_2 \longrightarrow 0 ,$$

and f to be the identity

$$id_X : X \longrightarrow X .$$

Let $g: X \longrightarrow 0$ be the zero map. Then $g \in Qis$, and $gf = 0$, but f is not homotopic to zero, as one sees easily.

2. If f satisfies (*), then f induces the zero map on cohomology, but the converse is false. For example take $f: X \longrightarrow Y$ as follows:

$$\begin{array}{llcccccccc} X: & & 0 & \longrightarrow & \mathbb{Z} & \xrightarrow{2} & \mathbb{Z} & \longrightarrow & 0 & \\ \downarrow f & & & & \downarrow 1 & & \downarrow 2 & & & \\ Y: & & 0 & \longrightarrow & \mathbb{Z} & \xrightarrow{1} & \mathbb{Z}_3 & \longrightarrow & 0 & . \end{array}$$

Now f induces the zero-map on cohomology, but there does not exist $t: X' \longrightarrow X$ in Qis, such that $ft \sim 0$. (The reader can supply this proof as follows: take a cycle $x \in X'$ such that $t(x)$ generates the single cohomology group $\mathbb{Z}_2$ of X. If k is a homotopy operator for ft, show that $2k(x) = 1$, which is impossible.)

So for two maps $f,g: X \longrightarrow Y$ of complexes, we see that the following implications are all strict:

$$f = g \Longrightarrow f \text{ homotopic to } g$$

$$\Longrightarrow f \text{ and } g \text{ give the same morphism in } D(A)$$

$$\Longrightarrow f \text{ and } g \text{ give the same map on cohomology.}$$

Proposition 4.3. The functor $A \longrightarrow D(A)$, which sends each object X of A into the complex consisting of X in degree 0, and 0 elsewhere, gives an equivalence of the category A with the full subcategory of $D(A)$ consisting of those complexes $X^\cdot$ such that $H^i(X^\cdot) = 0$ for $i \neq 0$.

Proof. Left to reader.

We will now give three lemmae, and another description of the derived category $D^+(A)$ when A has enough injectives.

Lemma 4.4. Let A be an abelian category, and let $f: Z^\cdot \longrightarrow I^\cdot$ be a morphism of complexes of objects of A. Assume

1) $Z^\cdot$ is acyclic

2) Each I^p is injective

3) $I^\cdot$ is bounded below.

Then f is homotopic to zero.

Proof. Well known (and easy).

Lemma 4.5. Let A be an abelian category and let $s\colon I^\cdot \longrightarrow Y^\cdot$ be a morphism of complexes of objects of A. Assume

1) s induces an isomorphism on cohomology

2) each I^p is injective

3) I^p is bounded below.

Then s has a homotopy inverse.

Proof. Suppose given $s\colon I^\cdot \longrightarrow Y^\cdot$ as above. Let $Z^\cdot = T(I^\cdot)\oplus Y^\cdot$ be the mapping cone of s. Then $Z^\cdot$ is acyclic, and so the triangular morphism $v\colon Z^\cdot \longrightarrow T(I^\cdot)$ satisfies the conditions of Lemma 4.4, and so is homotopic to zero. Let us call the homotopy operator

$$(k,t)\colon\ T(I^\cdot)\oplus Y^\cdot \longrightarrow I^\cdot .$$

Then we have the equation

$$v = (id_{I^\cdot}, 0) = (k,t)\, d_Z + d_I\, (k,t) .$$

Separating the components, we find

$$id_I = dk + kd + ts$$

and

$$dt - td = 0.$$

Thus $t: Y^\cdot \longrightarrow I^\cdot$ is a morphism of complexes, and id_I is homotopic to ts, so t is a homotopy inverse of s.

Lemma 4.6. Let A be an abelian category.

1). Let P be a subset of Ob A and assume

(i) Every object of A admits an injection into an element of P.

Then every $X^\cdot \in K^+(A)$ admits a quasi-isomorphism into a bounded below complex $I^\cdot$ of objects of P.

2). Assume furthermore that P satisfies

(ii) If $0 \longrightarrow X \longrightarrow Y \longrightarrow Z \longrightarrow 0$ is a short exact sequence, with $X \in P$, then $Y \in P \Longleftrightarrow Z \in P$.

(iii) There exists a positive integer n, such that if

$$X^0 \longrightarrow X^1 \longrightarrow \cdots \longrightarrow X^{n-1} \longrightarrow X^n \longrightarrow 0$$

is an exact sequence, and $X^0,\ldots,X^{n-1} \in P$, then $X^n \in P$.

Then every $X^\cdot \in K(A)$ admits a quasi-isomorphism into a complex $I^\cdot$ of objects of P.

3). Let A' be a thick subcategory of A, and assume that A' has enough A-injectives. Then every $X^\cdot \in K^+_{A'}(A)$ admits a quasi-isomorphism into a bounded below complex $I^\cdot$ of A-injective objects of A'.

<u>Proofs</u>. 1). We may assume $X^p = 0$ for $p < 0$. Embed $X^o \longrightarrow I^o$ with I^o in P. Having defined $I^o, I^1, \ldots, I^p$, choose I^{p+1} to be an element of P containing

$$I^p/\text{im}\, I^{p-1} \oplus_{X^p} X^{p+1} ,$$

and define the maps $I^p \longrightarrow I^{p+1}$ and $X^{p+1} \longrightarrow I^{p+1}$ in the obvious way. One checks easily that $X^{\cdot} \longrightarrow I^{\cdot}$ is a quasi-isomorphism. Note that in this construction all the maps $X^p \longrightarrow I^p$ are injective.

2). We proceed in several steps. Let $X^{\cdot}$ be a complex, and let i_o be an integer. Then by 1) we can find a quasi-isomorphism of the truncated complex

$$0 \longrightarrow 0 \longrightarrow X^{i_o} \longrightarrow X^{i_o+1} \longrightarrow \cdots$$

into a complex $I^{\cdot}$ of elements of P, with each $X^i \longrightarrow I^i$ injective. Define $X_o^{\cdot}$ to be the complex

$$\cdots \longrightarrow X^{i_o-2} \longrightarrow X^{i_o-1} \longrightarrow I^{i_o} \longrightarrow I^{i_o+1} \longrightarrow \cdots .$$

Then we have a quasi-isomorphism $X^{\cdot} \longrightarrow X_o^{\cdot}$ such that $X_o^i \in P$ for $i \geq i_o$, and each $X^i \longrightarrow X_o^i$ is injective.

Suppose given a complex $X_1^{\cdot}$ with $X_1^i \in P$ for $i \geq i_1$, and let $i_2 < i_1$. Then we will construct a quasi-isomorphism $X_1^{\cdot} \longrightarrow X_2^{\cdot}$ such that $X_2^i \in P$ for $i \geq i_2$, and $X_1^i = X_2^i$ for

$i \geq i_1+n$. (Here n is the integer of condition (iii) above.) Indeed, by the first step above, we can find a quasi-isomorphism $X_1^{\cdot} \longrightarrow X'^{\cdot}$ such that $X'^i \in P$ for $i \geq i_2$, and each $X_1^i \longrightarrow X'^i$ is injective. Let $Y^i = \operatorname{coker}(X_1^i \longrightarrow X'^i)$. Then $Y^{\cdot}$ is an acyclic complex, and $Y^i \in P$ for $i \geq i_1$, by condition (ii) above. Hence $B^i(Y^{\cdot}) \in P$ for $i \geq i_1+n$, by condition (iii). Now define $X_2^{\cdot}$ by

$$X_2^i = \begin{cases} X'^i & \text{for } i < i_1+n \\ B^i(X'^{\cdot}) \oplus_{X_1^{i-1}} X_1^i & \text{for } i = i_1+n \\ X_1^i & \text{for } i > i_1+n . \end{cases}$$

One sees easily that $X_1^{\cdot} \longrightarrow X_2^{\cdot}$ is a quasi-isomorphism. It follows from (ii) and the exact sequence

$$0 \longrightarrow X_1^i \longrightarrow B^i(X'^{\cdot}) \oplus_{X_1^{i-1}} X_1^i \longrightarrow B^i(Y^{\cdot}) \longrightarrow 0$$

that the middle term is in P, for $i \geq i_1+n$, so $X_2^{\cdot}$ is as required.

Now, given a complex $X^{\cdot} \in K(A)$, choose a sequence of integers $i_o > i_1 > \ldots$ tending to $-\infty$. Choose $X_o^{\cdot}$ for i_o as in the first step, and choose $X_1^{\cdot}, X_2^{\cdot}, \ldots$ for $i_1, i_2, \ldots$ successively as in the second step. Then we have quasi-isomorphisms

$X^{\cdot} \longrightarrow X^{\cdot}_{o} \longrightarrow X^{\cdot}_{1} \longrightarrow \ldots$ and for each i, the sequence $X^{i} \longrightarrow X^{i}_{o} \longrightarrow X^{i}_{1} \longrightarrow \ldots$ is eventually constant, and eventually in P. Hence $I^{\cdot} = \varinjlim X^{\cdot}_{r}$ is the required complex of objects of P.

3). We may assume $X^{i} = 0$ for $i < 0$. Embed $H^{o}(X^{\cdot})$ in I^{o}, an A-injective of A', which is possible since $H^{o}(X^{\cdot}) \in \text{Ob } A'$. Extend this to a map $f^{o}\colon X^{o} \longrightarrow I^{o}$, which is possible since I^{o} is A-injective. Having define $I^{o}, I^{1}, \ldots, I^{p}$, and $f^{i}\colon X^{i} \longrightarrow I^{i}$ for $i = 0, \ldots, p$, choose I^{p+1} to be an A-injective of A' containing

$$(*) \qquad I^{p}/\text{im } I^{p-1} \oplus_{X^{p}} Z^{p+1}(X^{\cdot}) .$$

We must check that this latter is in A'. Indeed, A' is a thick subcategory, so it is sufficient to note that $I^{p}/\text{im } I^{p-1} \in A'$ (one shows by induction that $B^{i}(I^{\cdot})$ and $Z^{i}(I^{\cdot})$ are in A' for all i), and that the quotient of (*) by $I^{p}/\text{im } I^{p-1}$ is $H^{p+1}(X^{\cdot})$, which is in A' by hypothesis. Extend the natural map $Z^{p+1}(X^{\cdot}) \longrightarrow I^{p+1}$ to a map $f^{p+1}\colon X^{p+1} \longrightarrow I^{p+1}$. One checks easily that the resulting map $f\colon X^{\cdot} \longrightarrow I^{\cdot}$ is a quasi-isomorphism, as required.

<u>Proposition 4.7</u>. Let A be an abelian category, and let I be the (additive) subcategory of injective objects of A. Then the natural functor

$$\alpha\colon\ K^{+}(I) \longrightarrow D^{+}(A)$$

is fully faithful. (Note that the results of section 3 carry over to additive subcategories of abelian categories.) Furthermore, if A has enough injectives (i.e., if every object of A admits an injection into an injective object) then α is an equivalence of categories.

<u>Proof</u>. We note that $K^{+}(I) \cap Qis$ is a multiplicative system in $K^{+}(I)$, by Proposition 4.2, and we observe by Lemma 4.5 that condition (ii) of Proposition 3.3 is satisfied for $K^{+}(I) \subseteq K^{+}(A)$ and Qis. Hence the natural functor

$$D^{+}(I) \longrightarrow D^{+}(A)$$

is fully faithful. But on the other hand, Lemma 4.5 shows also that every quasi-isomorphism in $K^{+}(I)$ is an isomorphism, hence $K^{+}(I) = D^{+}(I)$.

Now if A has enough injectives, applying Lemma 4.6 in the case $A = B$ and P = the injective objects, we see that every object of $D^+(A)$ is isomorphic to an object in $K^+(I)$, so α is an equivalence of categories.

Proposition 4.8. Let A be an abelian category, and let A' be a thick abelian subcategory. Assume that A' has enough A-injectives, i.e., every object of A' can be injected into an A-injective object of A'. Then the natural functor

$$D^+(A') \longrightarrow D^+_{A'}(A)$$

is an equivalence of categories.

Proof. We apply Proposition 3.3 to the inclusion $K^+(A') \to K^+(A)$. Clearly Qis is a multiplicative system in each. If $X^\cdot \to Y^\cdot$ is a quasi-isomorphism with $X^\cdot \in K^+(A')$, then $Y^\cdot$ has cohomology in A', and so by Lemma 4.6 admits a quasi-isomorphism $Y^\cdot \to I^\cdot$ with $I^\cdot \in K^+(A')$, each I^p injective in A. Hence condition (ii) is satisfied, and so the functor

$$D^+(A') \longrightarrow D^+(A)$$

is fully faithful. The same Lemma 4.6 also shows that the image is $D^+_{A'}(A)$.

<u>Exercise</u>. We leave to the reader the analogous statements of the last five results in the case of projective objects of A and $D^{-}(A)$.

§5. Derived Functors.

We will treat only the question of right derived covariant functors, leaving the reader to make the obvious modifications for left derived covariant functors, and right and left derived contravariant functors.

Let A,B be abelian categories, and let $F\colon\ K(A) \to K(B)$ be a ∂-functor (see §1). Such is the case, for example, if we are given an additive functor $F\colon A \to B$. It extends to $K(A)$.

In general, F will not take quasi-isomorphisms into quasi-isomorphisms—to say that it does is to say that it localizes and gives rise to a functor from $D(A)$ to $D(B)$. That will be the case, for example, if F is an exact functor.

Thus we are led to ask if there is a functor from $D(A)$ to $D(B)$ which is at least close to F, and this gives rise to the notion of derived functor below. Before giving the definition we generalize slightly.

Definition. Let A be an abelian category, and let $K^*(A)$ be a triangulated subcategory of $K(A)$. Note by Proposition 4.2 that $K^*(A) \cap \text{Qis}$ is a multiplicative system in $K^*(A)$. We say that $K^*(A)$ is a localizing subcategory of $K(A)$ if the natural functor

$$K^*(A)_{K^*(A)\cap Qis} \longrightarrow K(A)_{Qis} = D(A)$$

is fully faithful, and in that case we write $D^*(A)$ for the first of these categories.

<u>Examples</u>. 1. Any intersection of localizing subcategories is localizing.

2. $K^+(A)$, $K^-(A)$, and $K^b(A)$ are localizing subcategories of $K(A)$ (see section 4).

3. If A' is a thick subcategory of A, then $K_{A'}(A)$, $K^+_{A'}(A)$, $K^-_{A'}(A)$ and $K^b_{A'}(A)$ are localizing subcategories of $K(A)$ (see section 4).

4. The complexes of finite injective dimension form a localizing subcategory $K^+(A)_{fid}$ of $K(A)$ (see Corollary 7.7).

<u>Definition</u>. Let A and B be abelian categories, let $K^*(A)$ be a localizing subcategory of $K(A)$, and let

$$F: K^*(A) \longrightarrow K(B)$$

be a ∂-functor. Let Q denote the localization functor from $K^*(A)$ to $D^*(A)$ resp. $K(B)$ to $D(B)$. The <u>right derived functor</u> of F is a ∂-functor

$$\underline{\underline{R}}^*F\colon\ D^*(A) \longrightarrow D(B)$$

together with a morphism of functors

$$\xi\colon\ Q\cdot F \longrightarrow \underline{\underline{R}}^*F\cdot Q$$

from $K^*(A)$ to $D(B)$, with the following universal property: If

$$G\colon\ D^*(A) \longrightarrow D(B)$$

is any ∂-functor, and if

$$\zeta\colon\ Q\cdot F \longrightarrow G\cdot Q$$

is a morphism of functors, then there exists a unique morphism

$$\eta\colon\ \underline{\underline{R}}^*F \longrightarrow G$$

such that

$$\zeta = (\eta\cdot Q)\cdot\xi\ .$$

Remarks. 1. If $\underline{\underline{R}}^*F$ exists, it is unique up to unique isomorphism of functors.

2. If $K^*(A)$ is $K^+(A)$, $K^-(A)$, $K_{A'}(A)$, etc., we will write $\underline{\underline{R}}^+F$, $\underline{\underline{R}}^-F$, $\underline{\underline{R}}_{A'}F$ etc. for $\underline{\underline{R}}^*F$, and when no confusion can result, we will write simply $\underline{\underline{R}}F$ for all of these.

3. We will write R^iF for $H^i(\underline{\underline{R}}F)$, and it will follow from the results below that if F comes from a left-exact functor $F: A \to B$, and if A has enough injectives, then these are the usual derived functors of F.

4. If $\varphi: F \to G$ is a morphism of functors from $K^*(A)$ to $K(B)$, and if $\underline{\underline{R}}F$ and $\underline{\underline{R}}G$ both exist, then there is a unique morphism of functors

$$\underline{\underline{R}}\varphi: \underline{\underline{R}}F \longrightarrow \underline{\underline{R}}G$$

compatible with the ξ's. This follows immediately from the definition.

5. If $K^{**}(A) \subseteq K^*(A)$ are two localizing subcategories of $K(A)$, and if

$$F: K^*(A) \longrightarrow K(B)$$

is a ∂-functor, and if both $\underline{\underline{R}}^*F$ and $\underline{\underline{R}}^{**}(F|K^{**}(A))$ exist, then there is a natural morphism of functors

$$\underline{\underline{R}}^{**}(F|K^{**}(A)) \longrightarrow \underline{\underline{R}}^*F|D^{**}(A) .$$

We do not know if it is an isomorphism in general, but it will be in all the applications we have in mind (see e.g. Corollary 5.3 below).

Theorem 5.1. (Existence of derived functors). Let A, B, $K^*(A)$, and F be as in the definition above. Suppose there is a triangulated subcategory $L \subseteq K^*(A)$ such that

1) Every object of $K^*(A)$ admits a quasi-isomorphism into an object of L, and

2) If $I^\cdot \in \mathrm{Ob}\, L$ is acyclic (i.e., $H^i(I^\cdot) = 0$ for all i), Then $F(I^\cdot)$ is also acyclic.

Then F has a right derived functor $(\underline{\underline{R}}^*F, \xi)$. Furthermore, for any $I^\cdot \in \mathrm{Ob}\, L$, the map

$$\xi(I^\cdot): Q \cdot F(I^\cdot) \longrightarrow \underline{\underline{R}}^*F \cdot Q(I^\cdot)$$

is an isomorphism in $D(B)$.

Proof. First observe that the restriction of F to L takes quasi-isomorphisms into quasi-isomorphisms. Indeed, if $s: I_1^\cdot \longrightarrow I_2^\cdot$ is a quasi-isomorphism of objects of L, let $J^\cdot$ be the third side of a triangle built on s. Then $J^\cdot$ is acyclic, so $F(J^\cdot)$ is also, so $F(s)$ is a quasi-isomorphism. Hence F passes to the quotient to give a functor

$$\overline{F}: L_{Qis} \longrightarrow D(B)$$

with the property $\overline{F} \circ Q = Q \circ F$ on L. (We denote as usual by Q the morphism from a category to its localization.)

Second note that the hypotheses of Proposition 3.3, (ii) are satisfied for L, $K^*(A)$, and Qis, and so the natural functor

$$T: L_{Qis} \longrightarrow D^*(A)$$

is an equivalence of categories, using 1) above. Let U be a quasi-inverse of T, i.e., a functor

$$U: D^*(A) \longrightarrow L_{Qis}$$

together with functorial isomorphisms

$$\alpha: 1_{L_{Qis}} \longrightarrow U \circ T$$

and

$$\beta: 1_{D^*(A)} \longrightarrow T \circ U.$$

Then define

$$\underline{\underline{R}}^*F = \overline{F} \circ U .$$

We define a morphism of functors

$$\xi: Q \circ F \longrightarrow \underline{\underline{R}}^*F \circ Q = \overline{F} \circ U \circ Q$$

as follows. Let $X^{\cdot} \in \operatorname{Ob} K^*(A)$, and let $I^{\cdot} \in \operatorname{Ob} L$ be such that $Q(I^{\cdot}) = U \circ Q(X^{\cdot})$. We have an isomorphism in $D^*(A)$,

$$\beta(Q(X^\cdot)):\ Q(X^\cdot) \xrightarrow{\ \sim\ } T\circ U(Q(X^\cdot)) = T(QI^\cdot)\ .$$

This isomorphism can be represented by a diagram of morphisms

$$\begin{array}{ccccc} X^\cdot & & & & I^\cdot \\ & \overset{t}{\searrow} & & \overset{s}{\swarrow} & \\ & & Y^\cdot & & \end{array} \qquad (*)$$

in $K^*(A)$, with $Y^\cdot \in \mathrm{Ob}\, K^*(A)$ and $s,t \in \mathrm{Qis}$. Furthermore, by hypothesis 1) above, we may assume $Y^\cdot \in \mathrm{Ob}\, L$. Now applying the functor F, we get a diagram in $K(B)$

$$\begin{array}{ccccc} F(X^\cdot) & & & & F(I^\cdot) \\ & \overset{F(t)}{\searrow} & & \overset{F(s)}{\swarrow} & \\ & & F(Y^\cdot) & & \end{array}$$

where $F(s)$ is also a quasi-isomorphism, as we remarked above. This in turn gives a morphism in $D(B)$,

$$\xi(X^\cdot):\ Q\circ F(X^\cdot) \longrightarrow Q\circ F(I^\cdot) = \overline{F}\circ Q(I^\cdot) = \overline{F}\circ U\circ Q(X^\cdot) = \underline{\underline{R}}^*F\circ Q(X^\cdot).$$

One can now check without difficulty that $\xi(X^\cdot)$ does not depend on the choice of the diagram (*) above, that ξ gives a morphism of functors from $Q\circ F$ to $\underline{\underline{R}}^*F\circ Q$, and that the pair $(\underline{\underline{R}}^*F,\xi)$ is a derived functor of F.

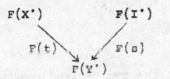

Now if $X^\cdot \in \mathrm{Ob}\, L$, then $F(t)$ in the construction above is also a quasi-isomorphism, and so $\xi(X^\cdot)$ is an isomorphism in $D(B)$, as required.

Proposition 5.2. Let A, B, $K^*(A)$, and F be as above, and let $K^{**}(A) \subseteq K^*(A)$ be another localizing subcategory of $K(A)$. Suppose there is a triangulated subcategory L of $K^*(A)$ satisfying hypotheses 1) and 2) of the theorem, and suppose, furthermore, that $L \cap K^{**}(A)$ satisfies 1) for $K^{**}(A)$. Then $\underline{\underline{R}}^*F$ and $\underline{\underline{R}}^{**}(F|K^{**}(A))$ both exist, and the natural map

$$\underline{\underline{R}}^{**}(F|K^{**}(A)) \longrightarrow \underline{\underline{R}}^*F|D^{**}(A)$$

is an isomorphism.

Proof. The existence of the two derived functors follows from the theorem. To prove the isomorphism, since every $X^\cdot \in \mathrm{Ob}\, D^{**}(A)$ is isomorphic to one coming from an object of L, we may assume that $X^\cdot = Q(I^\cdot)$ with $I^\cdot \in \mathrm{Ob}(L \cap K^{**}(A))$. Then the statement follows from the last part of the theorem.

Corollary 5.3. α. Let A,B be abelian categories, let $F\colon K^+(A) \to K(B)$ be a ∂-functor (defined for example by an additive functor $F_o\colon A \to B$), and assume that A has enough injectives. Then $\underline{\underline{R}}^+F$ exists.

β. Let A,B be abelian categories, let $F: A \longrightarrow B$ be an additive functor, and assume that there exists a subset P of Ob A having the properties (i) and (ii) of Lemma 4.6, and also

(iv) F carries short exact sequences of objects of P into short exact sequences.

Then $\underline{\underline{R}}^{+}F$ exists. (We denote also by F the extension of F to a ∂-functor $K^{+}(A) \longrightarrow K^{+}(B)$.)

γ. Let A,B be abelian categories, let $F: A \longrightarrow B$ be an additive functor, and assume that

a) The hypotheses of β above are satisfied, and

b) F has <u>finite cohomological dimension</u> on A, i.e., there is a positive integer n such that $R^{i}F(Y) = 0$ for all $Y \in \text{Ob } A$ and all $i > n$. (Note that $R^{i}F$ exists by β, so this makes sense.) Then $\underline{\underline{R}}F$ exists, and the restriction of $\underline{\underline{R}}F$ to $D^{+}(A)$ is equal to $\underline{\underline{R}}^{+}F$.

<u>Remark.</u> α is a special case of β, since if A has enough injectives, then the set P of injectives of A has properties (i),(ii), and (iv) for any additive functor F.

Proof. α. Let $L \subseteq K^+(A)$ be the triangulated subcategory of complexes of injective objects of A. Then by Lemma 4.6, 1), every $X^\bullet \in \mathrm{Ob}\, K^+(A)$ admits a quasi-isomorphism into an object of L. Furthermore, by Lemma 4.5, every quasi-isomorphism in L is an isomorphism. Hence condition 2) of the theorem is satisfied for any ∂-functor F, and we deduce that $\underline{\underline{R}}^+F$ exists.

β. In this case we take $L \subseteq K^+(A)$ to be the triangulated subcategory of complexes of objects of P. (Note L is a triangulated subcategory because P is stable under direct sums by (ii).) Condition 1) of the theorem is satisfied as above. For condition 2), let $Z^\bullet$ be acyclic. We use condition (ii) above repeatedly, and the fact that $Z^\bullet \in K^+(A)$, to show that $\ker d_Z^p \in P$ for every p. Then by condition (iv) of P, it follows that $F(Z^\bullet)$ is acyclic.

γ. We take P' to be the collection of F-acyclic objects of A, i.e., those $X \in \mathrm{Ob}\, A$ such that $R^iF(X) = 0$ for all $i > 0$. Then P' has properties (i),(ii) and (iii) of Lemma 4.6. We take $L \subseteq K(A)$ to be the complexes of objects of P'. Then using Lemma 4.6 and an argument similar to the one in β above, one sees that the hypotheses of the theorem are satisfied, so $\underline{\underline{R}}F$ exists.

One sees by Proposition 5.2 that the restriction of $\underline{\underline{R}}F$ to $D^{+}(A)$ is $\underline{\underline{R}}^{+}F$.

<u>Proposition 5.4</u>. Let A, B, C be abelian categories, let $K^{*}(A) \subseteq K(A)$ and $K^{\dagger}(B) \subseteq K(B)$ be localizing subcategories, and let

$$F\colon\ K^{*}(A) \longrightarrow K(B)$$

$$G\colon\ K^{\dagger}(B) \longrightarrow K(C)$$

be ∂-functors.

a). Assume that $F(K^{*}(A)) \subseteq K^{\dagger}(B)$, assume that $\underline{\underline{R}}^{*}F$, $\underline{\underline{R}}^{\dagger}G$, and $\underline{\underline{R}}^{*}(G \cdot F)$ exist, and assume that $\underline{\underline{R}}^{*}F(D^{*}(A)) \subseteq D^{\dagger}(B)$. Then there is a unique morphism of functors

$$\zeta\colon\ \underline{\underline{R}}^{*}(G \cdot F) \longrightarrow \underline{\underline{R}}^{\dagger}G \cdot \underline{\underline{R}}^{*}F$$

such that the diagram

$$\begin{array}{ccc} Q \cdot G \cdot F & \xrightarrow{\ \xi_G\ } & \underline{\underline{R}}^{\dagger}G \cdot Q \cdot F \\ \Big\downarrow{\scriptstyle \xi_{G \cdot F}} & & \Big\downarrow{\scriptstyle \xi_F} \\ \underline{\underline{R}}^{*}(G \cdot F) \cdot Q & \xrightarrow{\ \zeta \cdot Q\ } & \underline{\underline{R}}^{\dagger}G \cdot \underline{\underline{R}}^{*}F \cdot Q \end{array}$$

is commutative.

b). Assume that $F(K^*(A)) \subseteq K^{\dagger}(B)$, assume that there are triangulated subcategories $L \subseteq K^*(A)$ and $M \subseteq K^{\dagger}(B)$ satisfying the hypotheses of Theorem 5.1 for F and G, respectively, and assume that $F(L) \subseteq M$. Then the hypotheses of a) above are satisfied, and the morphism ζ which therefore exists, is an isomorphism.

<u>Proof</u>. Straightforward.

<u>Remarks</u>. 1. If F,G,H are three consecutive functors, then there is a commutative diagram of ζ's (provided they all exist):

$$\begin{array}{ccc}
\underline{\underline{R}}(H \circ G \circ F) & \xrightarrow{\;\xi_{G \circ F, H}\;} & \underline{\underline{R}}H \circ \underline{\underline{R}}(G \circ F) \\
\Big\downarrow{\scriptstyle \zeta_{F, H \circ G}} & & \Big\downarrow{\scriptstyle \zeta_{F,G}} \\
\underline{\underline{R}}(H \circ G) \circ \underline{\underline{R}}F & \longrightarrow & \underline{\underline{R}}H \circ \underline{\underline{R}}G \circ \underline{\underline{R}}F \;.
\end{array}$$

2. This proposition shows the convenience of derived functors in the context of derived categories. What used to be a spectral sequence becomes now simply a composition of functors. (And of course one can recover the old spectral sequence from this proposition by taking cohomology and using the spectral sequence of a double complex.)

Corollary 5.5. Left to the reader: Illustrate the Proposition in the style of Corollary 5.3.

Proposition 5.6. Let A be an abelian category, let A' be a thick abelian subcategory, let B be another abelian category, and let $F: K^+(A) \to K^+(B)$ be a ∂-functor. Suppose that $\underline{\underline{R}}^+F$ and $\underline{\underline{R}}^+(F|_{A'})$ both exist. Then there is a natural morphism

$$\zeta: \underline{\underline{R}}^+(F|_{A'}) \longrightarrow \varphi \circ \underline{\underline{R}}^+F$$

of functors from $D^+(A')$ to $D(B)$, where

$$\varphi: D^+(A') \longrightarrow D^+(A)$$

is the natural functor. If furthermore A' has enough A-injectives, and A has enough injectives, then ζ is an isomorphism.

Proof. The existence of ζ follows from the definition of the derived functor. If A' has enough A-injectives, and A has enough injectives, then we can use A-injectives to calculate both functors above, by Corollary 5.3a, and so ζ is an isomorphism. (Recall by Proposition 4.8 that φ is an equivalence of categories in that case.)

§6. Examples. Ext and $\underline{\underline{R}}$ Hom.

Definition. Let A be an abelian category, and let $X^{\cdot}, Y^{\cdot}$ be objects of $D(A)$. We define the ith hyperext of $X^{\cdot}, Y^{\cdot}$ to be

$$\mathrm{Ext}^{i}(X^{\cdot}, Y^{\cdot}) = \mathrm{Hom}_{D(A)}(X^{\cdot}, T^{i}(Y^{\cdot})).$$

Remarks. 1. If $X^{\cdot}, Y^{\cdot} \in D^{+}(A)$, then we get the same Ext by taking Hom in $D^{+}(A)$, for $D^{+}(A)$ is a full subcategory of $D(A)$.

2. This definition gives us in particular a definition of $\mathrm{Ext}^{i}(X,Y)$ for any $X,Y \in A$. We will see below that if A has enough injectives (so that the usual Ext is defined) then this definition agrees with the usual definition of Ext.

Proposition 6.1. Let

$$0 \longrightarrow X^{\cdot} \longrightarrow Y^{\cdot} \longrightarrow Z^{\cdot} \longrightarrow 0$$

be a short exact sequence of complexes of objects of A, and let $V^{\cdot}$ be another complex of objects of A. Then there are long exact sequences

$$\cdots \to \mathrm{Ext}^{i}(V^{\cdot}, X^{\cdot}) \to \mathrm{Ext}^{i}(V^{\cdot}, Y^{\cdot}) \to \mathrm{Ext}^{i}(V^{\cdot}, Z^{\cdot}) \to \mathrm{Ext}^{i+1}(V^{\cdot}, X^{\cdot}) \to \cdots$$

and

$$\cdots \to \mathrm{Ext}^{i}(Z^{\cdot}, V^{\cdot}) \to \mathrm{Ext}^{i}(Y^{\cdot}, V^{\cdot}) \to \mathrm{Ext}^{i}(X^{\cdot}, V^{\cdot}) \to \mathrm{Ext}^{i+1}(Z^{\cdot}, V^{\cdot}) \to \cdots$$

Proof. Let $W^\cdot$ be the third side of a triangle on $X^\cdot \to Y^\cdot$.

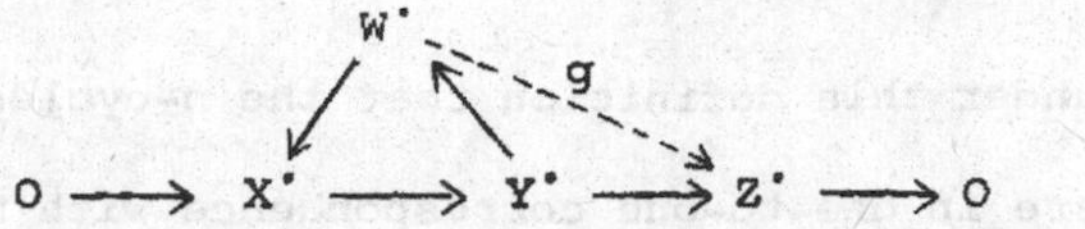

Then by Proposition 1.1b, there is a morphism of complexes $g\colon W^\cdot \to Z^\cdot$. Using the long exact sequence of cohomology of the short exact sequence, and of the triangle, and using the five-lemma, we see that g is a quasi-isomorphism, i.e., an isomorphism in $D(A)$. Hence we may replace $Z^\cdot$ by $W^\cdot$ in the conclusion, which then follows from the same proposition.

Remark. It follows from the proof that whenever

$$0 \to X^\cdot \to Y^\cdot \to Z^\cdot \to 0$$

is a short exact sequence of complexes of objects of A, then there is a morphism $Z^\cdot \to T(X^\cdot)$ in $D(A)$ making $X^\cdot, Y^\cdot, Z^\cdot$ into a triangle.

Now we will define a functor whose cohomology gives the Ext groups.

Definition. If $X^\cdot$ and $Y^\cdot$ are complexes of objects of A, we define a complex $\mathrm{Hom}^\cdot(X^\cdot, Y^\cdot)$ by

$$\mathrm{Hom}^n(X^\cdot, Y^\cdot) = \prod_{p\in\mathbb{Z}} \mathrm{Hom}_A(X^p, Y^{p+n})$$

and

$$d^n = \prod (d_X^{p-1} + (-1)^{n+1} d_Y^{p+n}) .$$

Notice under this definition that the n-cycles of the complex $Hom^\cdot(X^\cdot, Y^\cdot)$ are in one-to-one correspondence with morphisms of complexes of $X^\cdot$ to $T^n(Y^\cdot)$, and the n-boundaries correspond to those morphisms which are homotopic to zero. In other words,

$$H^n(Hom^\cdot(X^\cdot, Y^\cdot)) \cong Hom_{K(A)}(X^\cdot, T^n(Y^\cdot)) .$$

Now $Hom^\cdot$ is clearly a bi-∂-functor

$$Hom^\cdot : K(A)^\circ \times K(A) \longrightarrow K(Ab).$$

If A has enough injectives, we can calculate its derived functors.

<u>Lemma 6.2</u>. Let $X^\cdot \in Ob\, K(A)$ be a complex, and let $Y^\cdot \in Ob\, K^+(A)$ be a complex of injective objects. Assume either a) $Y^\cdot$ is acyclic, or b) $X^\cdot$ is acyclic. Then $Hom^\cdot(X^\cdot, Y^\cdot)$ is acyclic

<u>Proof</u>. By the remark above, one has only to check that any morphism of $X^\cdot$ to $T^n(Y^\cdot)$ is homotopic to zero, or, since $T^n(Y^\cdot)$ also satisfies the hypotheses of the lemma, it is enough to show that any morphism of $X^\cdot$ to $Y^\cdot$ is homotopic to zero. In case a), $Y^\cdot$ is split exact, and it is easy to construct the homotopy (left to reader). In case b) the result is Lemma 4.4.

Now suppose A has enough injectives, and let $L \subseteq K^+(A)$ be the triangulated subcategory of complexes of injective objects. Then using the lemma, part a), we see that for each $X^\cdot \in \text{Ob}\, K(A)$, L satisfies the hypotheses of Theorem 5.1 for the functor

$$\text{Hom}^\cdot(X^\cdot,\cdot)\colon \ K^+(A) \longrightarrow K(Ab).$$

Hence this functor has a right derived functor. It is easily seen to be functorial in $X^\cdot$, and so we have a bi-∂-functor

$$\underline{\underline{R}}_{II}\text{Hom}^\cdot\colon \ K(A)^\circ \times D^+(A) \longrightarrow D(Ab).$$

Now using the lemma, part b), we see that this functor is "exact" in the first variable, i.e., takes acyclic complexes into acyclic complexes, and hence passes to the quotient, giving a trivial right derived functor

$$\underline{\underline{R}}_{I}\underline{\underline{R}}_{II}\text{Hom}^\cdot\colon \ D(A)^\circ \times D^+(A) \longrightarrow D(Ab).$$

(We will denote this functor by $\underline{\underline{R}}\,\text{Hom}^\cdot$ when no confusion can result.)

Suppose on the other hand that A has enough projectives. Then by the usual process of "reversing the arrows" we see that there is also a functor

$$\underline{\underline{R}}_{II}\underline{\underline{R}}_{I}\text{ Hom}^\cdot\colon \ D^-(A)^\circ \times D(A) \longrightarrow D(A) \quad .$$

Now if A has enough injectives <u>and</u> enough projectives, then both functors $\underline{\underline{R}}_{I}\underline{\underline{R}}_{II}\text{Hom}^\cdot$ and $\underline{\underline{R}}_{II}\underline{\underline{R}}_{I}\text{Hom}^\cdot$ are defined on $D^-(A)^\circ \times D^+(A)$, and are canonically isomorphic, as we see by the lemma below. Thus we are justified in using the ambiguous notation $\underline{\underline{R}}\operatorname{Hom}^{\cdot}$.

<u>Lemma 6.3</u>. Let A, B, and C be abelian categories, and let

$$T: \; K^*(A) \times K^{\dagger}(B) \longrightarrow K(C)$$

be a bi-∂-functor. Suppose that $\underline{\underline{R}}_I\underline{\underline{R}}_{II}T$ and $\underline{\underline{R}}_{II}\underline{\underline{R}}_I T$ both exist (where the subscripts I,II denote the derived functor in the first or second variable, respectively). Then there is a unique isomorphism between them compatible with the morphisms $\xi_1: T \longrightarrow \underline{\underline{R}}_I\underline{\underline{R}}_{II}T$ and $\xi_2: \; T \longrightarrow \underline{\underline{R}}_{II}\underline{\underline{R}}_I T$.

<u>Proof</u>. Follows directly from the definition of derived functors.

<u>Theorem 6.4</u>. (Yoneda) Let A be an abelian category having enough injectives. Then for any $X^{\cdot} \in D(A)$, $Y^{\cdot} \in D^+(A)$,

$$H^i(\underline{\underline{R}}^+\operatorname{Hom}^{\cdot}(X^{\cdot},Y^{\cdot})) = \operatorname{Ext}^i(X^{\cdot},Y^{\cdot}).$$

<u>Corollary 6.5</u>. If A has enough injectives, then for any $X,Y \in A$, the $\operatorname{Ext}^i(X,Y)$ defined above is the usual Ext.

<u>Proof</u>. Let $I^{\cdot}$ be an injective resolution of $Y^{\cdot}$. Then using the above theorem, and Theorem 5.1, we have

$$\operatorname{Ext}^i(X,Y) = H^i(\operatorname{Hom}^{\cdot}(X,I^{\cdot})) ,$$

which is the usual definition.

Proof of Theorem 6.4. Let $s: Y^{\cdot} \longrightarrow I^{\cdot}$ be a quasi-isomorphism of $Y^{\cdot}$ into a complex of injectives $I^{\cdot}$. Then

$$\mathrm{Ext}^{i}(X^{\cdot},Y^{\cdot}) = \mathrm{Ext}^{i}(X^{\cdot},I^{\cdot}) = \mathrm{Hom}_{D(A)}(X^{\cdot},T^{i}(I^{\cdot})).$$

But by using Lemma 4.5, one sees that every morphism in $D(A)$ of a complex $X^{\cdot}$ to a complex of injectives bounded below, say $T^{i}(I^{\cdot})$, is represented by an actual morphism of complexes. Hence the above is equal to

$$\mathrm{Hom}_{K(A)}(X^{\cdot}, T^{i}(I^{\cdot})) = H^{i}(\mathrm{Hom}^{\cdot}(X^{\cdot},I^{\cdot}))$$

$$= H^{i}(\underline{\underline{R}}\,\mathrm{Hom}^{\cdot}(X^{\cdot},Y^{\cdot})) .$$

§7. Way-out functors and isomorphisms.

Definition. Let A and B be abelian categories, and let $F\colon D(A) \longrightarrow D(B)$ be a (covariant) ∂-functor. We say that F is way-out (right) if given $n_1 \in \mathbb{Z}$, there exists $n_2 \in \mathbb{Z}$ such that whenever $X^\cdot \in \mathrm{Ob}\, D(A)$ is a complex with $H^i(X^\cdot) = 0$ for all $i < n_2$, then $H^i(F(X^\cdot)) = 0$ for all $i < n_1$.

One defines similarly way-out left, way out in both directions. If $F\colon D(A) \longrightarrow D(B)$ is a contravariant ∂-functor, the definition of way-out right is the same, except that we reverse the inequality $i < n_2$ to be $i > n_2$.

Examples. 1. If $F_0\colon A \longrightarrow B$ is an additive functor satisfying the hypotheses of Corollary 5.3, α or β, then $\underline{\underline{R}}^+F$ is a way-out right functor. If F satisfies the hypotheses of γ, then $\underline{\underline{R}}F$ is way-out in both directions.

2. If $X^\cdot \in D(A)$ is an unbounded complex, then $\underline{\underline{R}}\,\mathrm{Hom}^\cdot(X^\cdot, Y^\cdot)$, for $Y^\cdot \in D^+(A)$, is in general not a way-out functor in $Y^\cdot$.

Proposition 7.1. (Lemma on Way-out Functors) Let A and B be abelian categories, let A' be a thick subcategory of A, let F and G be ∂-functors from $D^+_{A'}(A)$ (or $D_{A'}(A)$) to $D(B)$, and let $\eta\colon F \longrightarrow G$ be a morphism of functors.

(i) Assume that $\eta(X)$ is an isomorphism for all $X \in \text{Ob } A'$. Then $\eta(X^\cdot)$ is an isomorphism for all $X^\cdot \in \text{Ob } D^b_{A'}(A)$.

(ii) Assume that $\eta(X)$ is an isomorphism for all $X \in \text{Ob } A'$, and that F and G are both way-out right functors. Then $\eta(X^\cdot)$ is an isomorphism for all $X^\cdot \in D^+_{A'}(A)$.

(iii) Assume that $\eta(X)$ is an isomorphism for all $X \in \text{Ob } A'$, and that F and G are way-out in both directions. Then $\eta(X^\cdot)$ is an isomorphism for all $X^\cdot \in D_{A'}(A)$.

(iv) Let P be a subset of $\text{Ob } A'$ such that every object of A' admits an injection into an object of P. Assume $\eta(X)$ is an isomorphism for every $X \in P$, and that F and G are way-out right functors. Then $\eta(X)$ is an isomorphism for every $X \in \text{Ob } A'$.

<u>Definition</u>. Let $X^\cdot$ be a complex of objects of A. For an integer $n \in \mathbb{Z}$, we define the following truncations:

$$\tau_{>n}(X^\cdot): \quad \cdots \longrightarrow 0 \longrightarrow X^{n+1} \longrightarrow X^{n+2} \longrightarrow \cdots$$

$$\tau_{\leq n}(X^\cdot): \quad \cdots \longrightarrow X^{n-1} \longrightarrow X^{n} \longrightarrow 0 \longrightarrow \cdots$$

$$\sigma_{>n}(X^\cdot): \quad \cdots \longrightarrow 0 \longrightarrow \text{im } d^n \longrightarrow X^{n+1} \longrightarrow X^{n+2} \longrightarrow \cdots$$

$$\sigma_{\leq n}(X^\cdot): \quad \cdots \longrightarrow X^{n-1} \longrightarrow \ker d^n \longrightarrow 0 \longrightarrow \cdots \quad .$$

There are natural morphisms of complexes giving rise to the following exact sequences:

(1) $$0 \longrightarrow \tau_{>n}(X^\cdot) \longrightarrow X^\cdot \longrightarrow \tau_{\leq n}(X^\cdot) \longrightarrow 0$$

(2) $$0 \longrightarrow \sigma_{\leq n}(X^\cdot) \longrightarrow X^\cdot \longrightarrow \sigma_{>n}(X^\cdot) \longrightarrow 0 \ .$$

The truncation functor σ has the property

$$H^i(\sigma_{>n}(X^\cdot)) = \begin{cases} H^i(X^\cdot) & \text{for } i > n \\ 0 & \text{for } i \leq n \end{cases}$$

$$H^i(\sigma_{\leq n}(X^\cdot)) = \begin{cases} H^i(X^\cdot) & \text{for } i \leq n \\ 0 & \text{for } i > n \ . \end{cases}$$

<u>Lemma</u> 7.2. Let $X^\cdot$ be a complex of objects of A. Then there are triangles in $D(A)$

(3) $$\begin{array}{ccccc} & & \tau_{>n}(X^\cdot) & & \\ & \swarrow & & \nwarrow & \\ \tau_{\geq n}(X^\cdot) & & \longrightarrow & & X^n \end{array}$$

(4) $$\begin{array}{ccccc} & & \sigma_{>n}(X^\cdot) & & \\ & \swarrow & & \nwarrow & \\ H^n(X^\cdot) & & \longrightarrow & & \sigma_{\geq n}(X^\cdot) \end{array} \ .$$

Proof. The triangle (3) is deduced from an exact sequence of complexes

$$0 \longrightarrow \tau_{>n}(X^{\cdot}) \longrightarrow \tau_{\geq n}(X^{\cdot}) \longrightarrow X^n \longrightarrow 0$$

(see Remark following proof of Proposition 6.1). For the second, let $\sigma'_{\geq n}(X^{\cdot})$ be the complex

$$\cdots \longrightarrow 0 \longrightarrow X^n/\mathrm{im}\ d^{n-1} \longrightarrow X^{n+1} \longrightarrow \cdots .$$

Then there is a natural map $\sigma_{\geq n}(X^{\cdot}) \longrightarrow \sigma'_{\geq n}(X^{\cdot})$ which is a quasi-isomorphism, and there is an exact sequence of complexes

$$0 \longrightarrow H^n(X^{\cdot}) \longrightarrow \sigma'_{\geq n}(X^{\cdot}) \longrightarrow \sigma_{>n}(X^{\cdot}) \longrightarrow 0 .$$

Thus by the same remark we get a triangle in $D(A)$.

Proof of Proposition. (i) Let $X^{\cdot} \in \mathrm{Ob}\ D^b_{A'}(A)$. We prove, by descending induction on n, that

$$\eta(\sigma_{>n}(X^{\cdot})):\ F(\sigma_{>n}(X^{\cdot})) \longrightarrow G(\sigma_{>n}(X^{\cdot}))$$

is an isomorphism for all n. If n is large enough, then $\sigma_{>n}(X^{\cdot})$ has zero cohomology, since $X^{\cdot}$ has bounded cohomology. Hence it is the zero object of $D(A)$, and η of it is an isomorphism. The induction step follows from the hypotheses and Proposition 1.1c, applied to the triangle (4) above.

Now for n small enough, the natural map $X^{\cdot} \to \sigma_{>n}(X^{\cdot})$ is a quasi-isomorphism, hence an isomorphism in $D(A)$, and we are done.

(ii) Let $X^{\cdot} \in \mathrm{Ob}\, D^{+}_{A'}(A)$. To show $\eta(X^{\cdot})$ is an isomorphism, it is sufficient to show that

$$H^j(\eta(X^{\cdot})) : H^j(F(X^{\cdot})) \longrightarrow H^j(G(X^{\cdot}))$$

is an isomorphism for all $j \in \mathbb{Z}$. Given j, let $n_1 \geq j+2$, and choose n_2 as in the definition of way-out functors, to work for F and G both. From the exact sequence (2) above, we have a triangle in $D(A)$

$$\begin{array}{ccc} & \sigma_{>n_2}(X^{\cdot}) & \\ \swarrow & & \nwarrow \\ \sigma_{\leq n_2}(X^{\cdot}) & \longrightarrow & X^{\cdot} \end{array} .$$

Since $H^i(\sigma_{>n_2}(X^{\cdot})) = 0$ for $i \leq n_2$, we have

$$H^i\big(F(\sigma_{>n_2}(X^{\cdot}))\big) = 0 = H^i(G(\sigma_{>n_2}(X^{\cdot})))$$

for $i < n_1$, in particular for $i = j, j+1$. Therefore from the long exact sequence of cohomology we get isomorphisms

$$H^j\big(F(\sigma_{\leq n_2}(X^{\cdot}))\big) \xrightarrow{\sim} H^j(F(X^{\cdot}))$$

$$H^j\big(G(\sigma_{\leq n_2}(X^{\cdot}))\big) \xrightarrow{\sim} H^j(G(X^{\cdot})) .$$

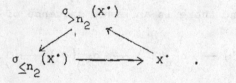

But $\sigma_{\leq n_2}(X^\cdot) \in D^b_{A'}(A)$, so η is an isomorphism on it by what we have just proved. Hence $H^j(\eta(X^\cdot))$ is an isomorphism, as required.

(iii) Given $X^\cdot \in D_{A'}(A)$, we treat first $\sigma_{\leq 0}(X^\cdot)$ and $\sigma_{>0}(X^\cdot)$ by case (ii), then glue the results by a triangle deduced from the exact sequence (2).

(iv) Given $X \in \operatorname{Ob} A'$, we can find by Lemma 4.6 a resolution $I^\cdot \in D^+_{A'}(A)$ of X, with each $I^p \in P$. Thus it is sufficient to show that $\eta(I^\cdot)$ is an isomorphism. This follows by applying the technique of (i) and (ii), but using the functors τ instead of σ.

q.e.d.

<u>Proposition 7.3</u>. Let A and B be abelian categories, and let $A' \subseteq A$ and $B' \subseteq B$ be thick abelian subcategories. Let F be a ∂-functor from $D^+_{A'}(A)$ (or $D_{A'}(A)$) to $D(B)$).

(i) Assume that $F(X) \in D_{B'}(B)$ for all $X \in \operatorname{Ob} A'$. Then $F(X^\cdot) \in D_{B'}(B)$ for all $X^\cdot \in D^b_{A'}(A)$.

(ii) Assume that $F(X) \in D_{B'}(B)$ for all $X \in \operatorname{Ob} A'$, and that F is a way-out right functor. Then $F(X^\cdot) \in D_{B'}(B)$ for all $X^\cdot \in D^+_{A'}(A)$.

(iii) As in (ii) but where we assume that F is way-out in both directions, and conclude that $F(X^\cdot) \in D_{B'}(B)$ for all $X^\cdot \in D_{A'}(A)$.

(iv) Let P be a subset of $\mathrm{Ob}\, A'$ such that every object of A' admits an injection into an object of P. Assume $F(X) \in D_{B'}(B)$ for every $X \in P$, and that F is a way-out right functor. Then $F(X) \in D_{B'}(B)$ for every $X \in \mathrm{Ob}\, A'$.

<u>Proof</u>. The proof is analogous to that of the previous proposition, and will be left to the reader.

<u>Proposition 7.4</u>. Let A and B be abelian categories, where A has enough injectives, and let $F: A \longrightarrow B$ be an additive functor which has cohomological dimension $\leq n$ on A. Let $P \subseteq \mathrm{Ob}\, A$ be the set of objects $X \in \mathrm{Ob}\, A$ such that $R^i F(X) = 0$ for all $i \neq n$, and assume that every object of A is a quotient of an element of P. Let $G = R^n F$. Then $\underline{\underline{R}}F$ and $\underline{\underline{L}}G$ exist, and there is a functorial isomorphism

$$\psi\colon \underline{\underline{R}}F \xrightarrow{\ \sim\ } \underline{\underline{L}}G[-n] ,$$

where "[-n]" means "shift n places to the right".

Proof. We know already from Corollary 5.3 that $\underline{\underline{R}}F$ exists. To see that the left derived functor $\underline{\underline{L}}G$ exists, we apply Theorem 5.1 with the arrows reversed. By hypothesis P satisfies condition (i)* of Lemma 4.6 (where * denotes reversal of arrows) and one sees easily from the definition of P that it satisfies also (ii)*, (iii)*, and (iv)* of Corollary 5.3. Therefore, if we let $L \subseteq K(A)$ be the triangulated subcategory consisting of complexes of elements of P, we see as in the proof of Corollary 5.3, γ, that the hypotheses 1)* and 2)* = 2) of the theorem are satisfied, and so $\underline{\underline{L}}G$ exists.

To define the morphism ψ, we observe, as in the proof of Theorem 5.1, that $T: L_{Qis} \to D(A)$ is an equivalence of categories, so it will be sufficient to define ψ on L_{Qis}. Or, using Proposition 3.4, it will be sufficient to define ψ on L itself. In other words, for every complex $X^{\cdot}$ of objects of P we must give a morphism in D(B),

$$\psi(X^{\cdot}): \quad \underline{\underline{R}}F(X^{\cdot}) \longrightarrow \underline{\underline{L}}G(X^{\cdot})[-n],$$

such that whenever $f: X^{\cdot} \to Y^{\cdot}$ is a morphism of such complexes, there is the usual commutative diagram: $\underline{\underline{L}}G(f) \circ \psi(X^{\cdot}) = \psi(Y^{\cdot}) \circ \underline{\underline{R}}F(f)$.

<u>Definition</u>. Let A be an abelian category, and $X^{\cdot}$ a complex of objects of A. We refer to [M, Ch. IV, §4] for the definition of double complexes, maps of double complexes, and homotopies of maps of double complexes. A <u>(right) Cartan-Eilenberg resolution</u> of $X^{\cdot}$ is a double complex $C^{\cdot\cdot}$ of injective objects of A, where $C^{p,q} = 0$ if $p < 0$, and an augmentation $\epsilon: X^{\cdot} \to C^{\cdot 0}$, such that for every $p \in \mathbb{Z}$, the maps

$$B^p(\epsilon): \quad B^p(X^{\cdot}) \longrightarrow B_I^{p\cdot}(C^{\cdot\cdot})$$

$$H^p(\epsilon): \quad H^p(X^{\cdot}) \longrightarrow H_I^{p\cdot}(C^{\cdot\cdot})$$

are injective resolutions in the ordinary sense (cf. [M, Ch. XVII, §1] where this is called an injective resolution of $X^{\cdot}$).

We recall for convenience some properties of double complexes and Cartan-Eilenberg resolutions.

<u>Lemma 7.5</u>. a) If A has enough injectives, then every complex $X^{\cdot}$ of objects of A has a Cartan-Eilenberg resolution.

b) If $f: X^{\cdot} \longrightarrow Y^{\cdot}$ is a map of complexes, and if $C^{\cdot\cdot}, D^{\cdot\cdot}$ are Cartan-Eilenberg resolutions of $X^{\cdot}$ and $Y^{\cdot}$, respectively, then there is a map $F: C^{\cdot\cdot} \longrightarrow D^{\cdot\cdot}$ of double complexes lying over f.

c) If $f,g: X^{\cdot} \longrightarrow Y^{\cdot}$ are homotopic maps, and $F,G: C^{\cdot\cdot} \longrightarrow D^{\cdot\cdot}$ are maps of Cartan-Eilenberg resolutions lying over f and g, respectively, then F is homotopic to G.

d) If $F,G: C^{\cdot\cdot} \longrightarrow D^{\cdot\cdot}$ are homotopic maps of double complexes, then $s(F),s(G): s(C^{\cdot\cdot}) \rightarrow s(D^{\cdot\cdot})$ are homotopic maps of the associated simple complexes.

e) If $F,G: C^{\cdot\cdot} \longrightarrow D^{\cdot\cdot}$ are homotopic maps of double complexes, and if we define truncation $\sigma^{II}_{\leq n}$ as above, with respect to d_2, then the restrictions of F and G to be maps $\sigma^{II}_{\leq n}(C^{\cdot\cdot}) \longrightarrow \sigma^{II}_{\leq n}(D^{\cdot\cdot})$ are homotopic.

f) If $\epsilon: X^{\cdot} \longrightarrow C^{\cdot\cdot}$ is an augmentation of $X^{\cdot}$ into a double complex $C^{\cdot\cdot}$ with $C^{pq} = 0$ for $q < 0$ and $q > n_o$ for suitable n_o, and such that for each p, $X^p \longrightarrow C^{p\cdot}$ is a resolution, then the natural map $X^{\cdot} \longrightarrow s(C^{\cdot\cdot})$ into the associated simple complex is a quasi-isomorphism.

<u>Proofs</u>. a), b), and c) are in [M, Ch. XVII, Prop. 1.2]. d) is in [M, Ch. IV, §4]. e) is easy, and f) follows either from the spectral sequence of a double complex [EGA 0_{III} 11.3.3 (ii)] or by an easy direct calculation.

<u>Proof of Proposition, continued</u>. Given $X^{\cdot} \in \operatorname{Ob} L$, let $C^{\cdot\cdot}$ be a Cartan-Eilenberg resolution of $X^{\cdot}$, which exists since A has enough injectives. Let $C'^{\cdot\cdot}$ be the truncated complex $\sigma^{II}_{\leq n}(C^{\cdot\cdot})$, i.e.,

$$C'^{pq} = \begin{cases} C^{pq} & \text{for } q < n \\ \ker d_2^{pn} & \text{for } q = n \\ 0 & \text{for } q > n . \end{cases}$$

Then for each $p \in \mathbb{Z}$ we have an exact sequence

$$0 \longrightarrow X^p \longrightarrow C'^{p0} \longrightarrow C'^{p1} \longrightarrow \cdots \longrightarrow C'^{p,n-1} \longrightarrow C'^{pn} \longrightarrow 0 .$$

Now $C'^{p0},\ldots,C'^{p,n-1}$ are all injective, and $X^p \in P$, so C'^{pn} is F-acyclic. Hence this resolution may be used to calculate $\underline{\underline{R}}F(X^p)$ (see end of Theorem 5.1) and we have an isomorphism

$$\varphi(t^p):\ G(X^p) = R^nF(X^p) \xrightarrow{\ \sim\ } F(C'^{pn})/\operatorname{Im} F(C'^{p,n-1}) ,$$

where t^p is the quasi-isomorphism $X^p \longrightarrow C'^{p\cdot}$. This can be used to construct a map

$$\alpha^p:\ F(C'^{pn}) \longrightarrow G(X^p)$$

whence a map

$$\alpha:\ F(C'^{\cdot\cdot}) \longrightarrow G(X^{\cdot})$$

of double complexes, where the second is concentrated in the nth row. Taking associated simple complexes we have

$$s(\alpha):\ F(sC'^{\cdot\cdot}) \longrightarrow G(X^{\cdot})[-n].$$

But now $u: X^{\cdot} \longrightarrow sC'^{\cdot\cdot}$ is a quasi-isomorphism into a complex of F-acyclic objects, by the lemma, part f), and so there is an isomorphism

$$\varphi(u):\quad \underline{\underline{R}}F(X^{\cdot}) \longrightarrow F(sC'^{\cdot\cdot}) .$$

But also $X^\cdot$ is made of G-acyclic objects, so there is an isomorphism

$$\varphi(id_{X^\cdot}): \; G(X^\cdot) \longrightarrow \underline{\underline{L}}G(X^\cdot)$$

and composing we can define a morphism

$$\psi(X^\cdot): \quad \underline{\underline{R}}F(X^\cdot) \longrightarrow \underline{\underline{L}}G(X^\cdot).$$

I do not care whether $\psi(X^\cdot)$ depends on the choice of the Cartan-Eilenberg resolution $C^{\cdot\cdot}$. It will be sufficient to verify that whenever $f: X^\cdot \rightarrow Y^\cdot$ is a morphism of complexes in L, then $\psi(X^\cdot)$ and $\psi(Y^\cdot)$ defined as above, fit into a commutative diagram. This is not hard to show using the results of the lemma above, and can safely be left to the reader.

Thus we have a morphism of functors

$$\psi: \quad \underline{\underline{R}}F \longrightarrow \underline{\underline{L}}G[-n] .$$

To show that ψ is an isomorphism, we note that $\underline{\underline{R}}F$ and $\underline{\underline{L}}G$ are way-out in both directions, and so, by the lemma on way-out functors, we reduce to showing $\psi(X)$ is an isomorphism for any $X \in P$. But that is clear from the construction. q.e.d.

<u>Proposition 7.6</u>. Let A be an abelian category with enough injectives, and let $X^\cdot \in \mathrm{Ob}\, K^+(A)$. Then the following conditions are equivalent:

(i) $X^\cdot$ admits a quasi-isomorphism $X^\cdot \to I^\cdot$ into a bounded complex of injective objects of A.

(ii) The functor $F = \underline{\underline{R}}\,\mathrm{Hom}^\cdot(\cdot, X^\cdot)$ from $D(A)^\circ$ to $D(Ab)$ is way-out left (and hence way-out in both directions).

(iii) There is an integer n_o such that $\mathrm{Ext}^i(Y, X^\cdot) = 0$ for all $Y \in \mathrm{Ob}\, A$ and all $i > n_o$.

<u>Proof</u>. (i) $\Rightarrow$ (ii) We may assume that $X^\cdot$ is a bounded complex of injectives, and calculate $F(Y^\cdot)$ as $\mathrm{Hom}^\cdot(Y^\cdot, X^\cdot)$. Then it is clear that F is way-out in both directions.

(ii) $\Rightarrow$ (iii) Choose n_o such that whenever $Y^\cdot \in D(A)$ and $H^i(Y^\cdot) = 0$ for $i < n_o$, then $H^i(F(Y^\cdot)) = 0$ for $i > 0$. (This is possible since F is way-out left.) Then this n_o will do. Indeed, let $Y \in \mathrm{Ob}\, A$, and let $Y'^\cdot$ be the complex $T^{-n_o}(Y)$. Then $H^i(Y'^\cdot) = 0$ for $i < n_o$, so $H^i(F(Y'^\cdot)) = 0$ for $i > 0$. But $H^i(F(Y'^\cdot)) = H^i(F(T^{-n_o}(Y))) = H^i(T^{n_o}F(Y)) = H^{i+n_o}(F(Y))$. This says $\mathrm{Ext}^j(Y, X^\cdot) = 0$ for $j > n_o$, as required.

(iii) $\Rightarrow$ (i). Let $X^\cdot$ satisfy condition (iii) for n_o. Since A has enough injectives, we can find an injective resolution of $X^\cdot$

(Lemma 4.6), i.e., a quasi-isomorphism $s\colon X^{\cdot} \to X'^{\cdot}$ where $X'^{\cdot}$ is a complex of injective objects of A, bounded below.

I claim that $H^i(X'^{\cdot}) = 0$ for $i > n_o$. Suppose to the contrary that $H^m(X'^{\cdot}) \neq 0$ for some $m > n_o$. Then from the exact sequence

$$0 \to B^m(X'^{\cdot}) \to Z^m(X'^{\cdot}) \to H^m(X'^{\cdot}) \to 0$$

we see that the first inclusion is strict. Hence there exists a $Y \in \mathrm{Ob}\, A$ (e.g., $Y = Z^m(X'^{\cdot})$) such that the inclusion

$$\mathrm{Hom}(Y, B^m(X'^{\cdot})) \to \mathrm{Hom}(Y, Z^m(X'^{\cdot}))$$

is strict. However, there is a commutative diagram

$$\begin{array}{ccc} B^m(\mathrm{Hom}^{\cdot}(Y,X'^{\cdot})) & \longrightarrow & Z^m(\mathrm{Hom}^{\cdot}(Y,X'^{\cdot})) \\ \downarrow & & \downarrow \approx \\ \mathrm{Hom}(Y, B^m(X'^{\cdot})) & \longrightarrow & \mathrm{Hom}(Y, Z^m(X'^{\cdot})) \end{array}$$

with an isomorphism on the right since Hom is left exact. But by hypothesis $\mathrm{Ext}^m(Y,X^{\cdot}) = H^m(\mathrm{Hom}^{\cdot}(Y,X'^{\cdot})) = 0$, so the arrow on top is an isomorphism, which gives a contradiction. Hence $H^i(X'^{\cdot}) = 0$ for $i > n_o$ as claimed.

It follows that the map

$$i\colon \sigma_{\leq n}(X'^{\cdot}) \to X'^{\cdot}$$

is a quasi-isomorphism for any $n \geq n_o$ (using the notation introduced earlier in this section). Now I claim that $Z^{n+1}(X^{\cdot}) = B^{n+1}(X^{\cdot})$ is injective for any $n \geq n_o$. This will show that $\sigma_{\leq n+1}(X^{\cdot})$ is a bounded complex of injectives.

Consider the exact sequence of complexes

$$0 \longrightarrow \sigma_{\leq n}(X^{\cdot}) \longrightarrow \tau_{\leq n}(X^{\cdot}) \longrightarrow B^{n+1}(X^{\cdot}[-n] \longrightarrow 0$$

for $n \geq n_o$. We will show that $B^{n+1} = B^{n+1}(X^{\cdot})$ is injective by showing that $\mathrm{Ext}^1(Y, B^{n+1}) = 0$ for all $Y \in \mathrm{Ob}\, A$. For any Y, we have

$$\mathrm{Ext}^1(Y, B^{n+1}) = \mathrm{Ext}^{n+1}(Y, B^{n+1}[-n])$$

and there is an exact sequence

$$\cdots \longrightarrow \mathrm{Ext}^{n+1}(Y, \tau_{\leq n}(X^{\cdot})) \longrightarrow \mathrm{Ext}^{n+1}(Y, B^{n+1}[-n]) \longrightarrow \mathrm{Ext}^{n+2}(Y, \sigma_{\leq n}(X^{\cdot})) \longrightarrow \cdot$$

The group on the left is zero since $\tau_{\leq n}(X^{\cdot})$ is an injective complex which stops in degree n. The group on the right is zero by our hypothesis and the fact that $\sigma_{\leq n}(X^{\cdot})$ is isomorphic to $X^{\cdot}$ in $D^+(A)$. Hence the middle group is zero.

Now for $n \geq n_o+1$ we have seen that $\sigma_{\leq n}(X^{\cdot})$ is a bounded complex of injectives, and that the map

$$i: \sigma_{\leq n}(X'^{\cdot}) \longrightarrow X'^{\cdot}$$

is a quasi-isomorphism. By Lemma 4.5, i has a homotopy inverse, say f. Therefore the composition

$$fs: X^{\cdot} \longrightarrow \sigma_{\leq n}(X'^{\cdot})$$

gives the required quasi-isomorphism of $X^{\cdot}$ into a bounded complex of injectives.

Definition. If $X^{\cdot} \in \text{Ob } K^{+}(A)$ satisfies the equivalent conditions of the Proposition, we say that $X^{\cdot}$ has finite injective dimension.

Corollary 7.7. The complexes $X^{\cdot} \in K^{+}(A)$ with finite injective dimension form a localizing (see §5) triangulated subcategory of $K(A)$.

Proof. The fact that they form a triangulated subcategory follows e.g. from condition (iii) and the long exact sequence of Ext. We see that the subcategory is localizing by applying Proposition 3.3(i), since anything in $K^{+}(A)$ quasi-isomorphic to an element of the subcategory is already in the subcategory, and $K^{+}(A)$ is a localizing subcategory of $K(A)$.

<u>Definition</u>. We denote the category of $X^\cdot \in K^+(A)$ with finite injective dimension by $K^+(A)_{fid}$. The corresponding subcategory of $D^+(A)$ is denoted by $D^+(A)_{fid}$.

<u>Remark</u>. These subcategories are essentially different from the subcategories of the form $K_{A'}(A)$ and $D_{A'}(A)$ we have considered before, because they cannot be characterized by putting conditions on the cohomology objects $H^i(X^\cdot)$.

CHAPTER II. APPLICATION TO PRESCHEMES

In this chapter we will discuss the various categories of sheaves we plan to use in the sequel, their derived categories, various functors between them, and relations among these functors. In §7 we give the structure of injective sheaves of $\mathcal{O}_X$-modules on a locally noetherian prescheme X.

§1. Categories of sheaves.

Let $(X, \mathcal{O}_X)$ be a prescheme. We will denote by $\mathrm{Mod}(X)$ the category of sheaves of $\mathcal{O}_X$-modules, and we will denote its derived category by $D(X) = D(\mathrm{Mod}(X))$.

We will consider the thick subcategories of $\mathrm{Mod}(X)$ consisting of the quasi-coherent sheaves, denoted by $\mathrm{Qco}(X)$, and the coherent sheaves, denoted by $\mathrm{Coh}(X)$. (Cf. [EGA 0, §5] for definitions and the fact that these are thick subcategories.) We will write $D_{qc}(X)$ for $D_{\mathrm{Qco}(X)}(\mathrm{Mod}(X))$ and $D_c(X)$ for $D_{\mathrm{Coh}(X)}(\mathrm{Mod}(X))$, using the notation of [I.4]. Similarly $D_{qc}^+(X)$ etc.

<u>Proposition 1.1.</u> There are enough injectives in the categories $\mathrm{Mod}(X)$ and $\mathrm{Qco}(X)$.

<u>Proof</u>. This is well known, e.g., [T, 1.10.1], or, for Mod(X), [G, II 7.1.1].

<u>Proposition 1.2</u>. Every sheaf of $\mathcal{O}_X$-modules is a quotient of a flat $\mathcal{O}_X$-module.

<u>Proof</u>. If U is an open subset of X, we denote by $\mathcal{O}_{X,U}$ the sheaf which is $\mathcal{O}_X$ on U and 0 outside [G, II Thm. 2.9.1]. Then $\mathcal{O}_{X,U}$ is a flat $\mathcal{O}_X$-module, and any direct sum of these is also flat. If F is any sheaf of $\mathcal{O}_X$-modules, let $s_i \in \Gamma(U_i, F)$ be a set of generators of F as an $\mathcal{O}_X$-module. Then F is a quotient of $\Sigma\, \mathcal{O}_{X,U_i}$, by sending $1 \in \Gamma(U_i, \mathcal{O}_{X,U_i})$ to s_i.

§2. The derived functors of f_* and Γ.

Let $f\colon X \longrightarrow Y$ be a morphism of preschemes. Then the functor f_* sends $Mod(X)$ to $Mod(Y)$. Since the first of these categories has enough injectives, we have a derived functor

$$\underline{\underline{R}}^+ f_*\colon D^+(X) \longrightarrow D^+(Y).$$

There are two cases in which we can define $\underline{\underline{R}}f_*$ for unbounded complexes. If X is noetherian of finite Krull dimension n, then $H^i(X,F) = 0$ for $i > n$ and every sheaf F of abelian groups on X [T 3.6.5]. Note that if F is an $\mathcal{O}_X$-module, then its cohomology as an $\mathcal{O}_X$-module is the same as its cohomology as an abelian sheaf [EGA 0_{III} 12.1.1]. Therefore we have $R^i f_*(F) = 0$ for $i > n$, since this sheaf is the sheaf associated to the presheaf $V \longrightarrow H^i(f^{-1}(V),F)$. Thus f_* has finite cohomological dimension on $Mod(X)$, and using [I 5.3γ] we can define

$$\underline{\underline{R}}f_*\colon D(X) \longrightarrow D(Y).$$

On the other hand, if f_* has finite cohomological dimension on $Qco(X)$, we can define

$$\underline{\underline{R}}f_*\colon D(Qco(X)) \longrightarrow D(Qco(Y)) .$$

This will be the case if

a) f is separated, quasi-compact, and Y is quasi-compact, by [EGA III 1.4.12], or

b) if X,Y are locally noetherian, f is of finite type, and the fibres of f are of bounded dimension (use [EGA III 4.1.5] and [T.3.6.5]).

We will often make statements in only one of these two contexts, and will let the reader make the necessary modifications for the other. We will also write $\underline{\underline{R}}f_*$ instead of $\underline{\underline{R}}^+f_*$, when no confusion can arise.

Proposition 2.1. If $f\colon X \longrightarrow Y$ is separated and quasi-compact, then $\underline{\underline{R}}f_*$ takes $D^+_{qc}(X)$ into $D^+_{qc}(Y)$. If, furthermore, X is noetherian of finite Krull dimension, $\underline{\underline{R}}f_*$ takes $D_{qc}(X)$ into $D_{qc}(Y)$.

Proof. By [EGA III 1.4.10], if F is quasi-coherent on X, then all the $R^if_*(F)$ are quasi-coherent sheaves on Y. In other words, $\underline{\underline{R}}f_*(F) \in D^+_{qc}(Y)$. But $\underline{\underline{R}}f_*$ is a way-out functor, so the result follows from [I.7.3].

Proposition 2.2. If $f: X \longrightarrow Y$ is a proper morphism, and Y is locally noetherian, then $\underline{\underline{R}}f_*$ sends $D_c^+(X)$ into $D_c^+(Y)$. If, furthermore, X is noetherian of finite Krull dimension, $\underline{\underline{R}}f_*$ takes $D_c(X)$ into $D_c(Y)$.

Proof. Same as previous proposition, using [EGA III 3.2.1].

If X is a prescheme, the global section functor Γ has a right derived functor

$$\underline{\underline{R}}^+\Gamma: \quad D^+(X) \longrightarrow D(Ab) .$$

As above, we can extend the domain of definition of $\underline{\underline{R}}\Gamma$ to $D(X)$ if X is noetherian of finite Krull dimension.

§3. The derived functor of $\underline{Hom}^\bullet$.

Let X be a prescheme, and let $F^\bullet, G^\bullet \in K(Mod(X))$. Then we define $\underline{Hom}^\bullet(F^\bullet, G^\bullet) \in K(Mod(X))$ as in [I.6] by

$$\underline{Hom}^n(F^\bullet, G^\bullet) = \prod_{p\in\mathbb{Z}} \underline{Hom}(F^p, G^{p+n})$$

and

$$d = d_F + (-1)^{n+1} d_G.$$

Then $\underline{Hom}^\bullet$ is a bi-∂-functor

$$\underline{Hom}^\bullet : K(Mod(X))^\circ \times K(Mod(X)) \longrightarrow K(Mod(X)).$$

<u>Lemma 3.1</u>. Let $F^\bullet$ be a complex of $\mathcal{O}_X$-modules, and let $G^\bullet$ be a complex of injective $\mathcal{O}_X$-modules, bounded below. Assume that either $F^\bullet$ or $G^\bullet$ is acyclic. Then $\underline{Hom}^\bullet(F^\bullet, G^\bullet)$ is acyclic.

<u>Proof</u>. It is enough to show for each open subset U of X that $\Gamma(U, \underline{Hom}^\bullet(F^\bullet, G^\bullet)) = Hom^\bullet(F^\bullet|U, G^\bullet|U)$ is acyclic. But this follows from [I.6.2] applied to the abelian category of $\mathcal{O}_U$-modules, since if G is an injective $\mathcal{O}_X$-module, then $G|U$ is an injective $\mathcal{O}_U$-module.

Using this Lemma, we can follow the method of [I.6] and obtain a derived functor

$$\underline{\underline{R}}\ \underline{Hom}^{\cdot}: \quad D(X)^{\circ} \times D^{+}(X) \longrightarrow D(X)$$

(more precisely $\underline{\underline{R}}_{I}\underline{\underline{R}}_{II}\ \underline{Hom}^{\cdot}$).

Exercise. Suppose that X is locally noetherian, and that every coherent $\mathcal{O}_X$-module is a quotient of a locally free $\mathcal{O}_X$-module of finite rank (e.g. X quasi-projective over a field). Show then that

$$\underline{\underline{R}}_{II}\underline{\underline{R}}_{I}\ \underline{Hom}^{\cdot}: \quad D_c^{-}(X)^{\circ} \times D(X) \longrightarrow D(X)$$

exists.

Problem. Without the hypotheses of the Exercise, study sheaves on X which are acyclic for $\underline{Hom}$ in the first variable. Find out whether there are enough of them, and hence whether $\underline{\underline{R}}_{I}\underline{Hom}^{\cdot}$ exists.

Definition. If $F^{\cdot} \in D(X)$ and $G^{\cdot} \in D^{+}(X)$, we define the local hyperext

$$\underline{Ext}^{i}(F^{\cdot},G^{\cdot}) = H^{i}(\underline{\underline{R}}\ \underline{Hom}^{\cdot}(F^{\cdot},G^{\cdot})) .$$

Lemma 3.2. Let X be a locally noetherian prescheme. Let F be a coherent $\mathcal{O}_X$-module, and let G be a coherent (resp. quasi-coherent) $\mathcal{O}_X$-module. Then for all $i \geq 0$ $\underline{\mathrm{Ext}}^i(F,G)$ is coherent (resp. quasi-coherent).

Proof. The coherent case is [EGA 0_{III} 12.3.3], and the quasi-coherent case is proved similarly.

Proposition 3.3. Let X be a locally noetherian prescheme. Let $F^\cdot \in D_c(X)$ and $G^\cdot \in D_c^+(X)$ (resp. $D_{qc}^+(X)$). Assume either

a) $F^\cdot \in D_c^-(X)$ or

b) $G^\cdot \in D_c^+(X)_{fid}$ (cf. [I.7]).

Then $\underline{\underline{R}}\,\underline{\mathrm{Hom}}^\cdot(F^\cdot,G^\cdot) \in D_c(X)$ (resp. $D_{qc}(X)$).

Proof. This follows from the lemma and [I.7.3] applied twice. The details are left to the reader.

§4. The derived functors of ⊗ and f^*.

Let X be a prescheme, and let $F^\cdot, G^\cdot \in K(\mathrm{Mod}(X))$. Then we define the tensor product $F^\cdot \otimes G^\cdot$ to be the simple complex associated to the double complex $(F^p \otimes G^q)$, i.e.,

$$(F^\cdot \otimes G^\cdot)^n = \sum_{p+q=n} F^p \otimes G^q$$

and

$$d = d_F + (-1)^n d_G .$$

Homotopies carry over to the tensor product, so we have a functor

$$\otimes: K(\mathrm{Mod}(X)) \times K(\mathrm{Mod}(X)) \longrightarrow K(\mathrm{Mod}(X)).$$

Lemma 4.1. Let $F^\cdot$ be a complex of $\mathcal{O}_X$-modules, and let $G^\cdot$ be a complex of flat $\mathcal{O}_X$-modules, bounded above. Assume that either

a) $G^\cdot$ is acyclic, or

b) $F^\cdot$ is acyclic,

and assume also that either

1) $F^\cdot$ is bounded above, or

2) $G^\cdot$ is bounded in both directions.

Then $F^\cdot \otimes G^\cdot$ is acyclic.

<u>Proof</u>. Let $K^{\cdot\cdot}$ be the double complex $K^{pq} = F^p \otimes G^q$. Then there are spectral sequences [EGA 0_{III} 11.3.2]

$$'E_2^{pq} = H_I^p H_{II}^q (K^{\cdot\cdot}) \Longrightarrow E^n = H^n(F^\cdot \otimes G^\cdot)$$

and
$$''E_2^{pq} = H_{II}^p H_I^q (K^{\cdot\cdot}) \Longrightarrow E^n = H^n(F^\cdot \otimes G^\cdot).$$

Our hypotheses 1) or 2) imply that these spectral sequences are biregular. In case a), one shows that $B^q(G^\cdot) = Z^q(G^\cdot)$ is flat for each q, and so $F^p \otimes G^\cdot$ is acyclic for each p. This implies that $'E_2^{pq} = 0$ for all p,q, and hence $E^n = 0$, and $F^\cdot \otimes G^\cdot$ is acyclic. In case b), $F^\cdot \otimes G^q$ is acyclic for each q, since G^q is flat, and so $''E_2^{pq} = 0$ for each p,q. Hence again $E^n = 0$, and $F^\cdot \otimes G^\cdot$ is acyclic.

Now let $F^\cdot \in K^-(\text{Mod}(X))$, and let $L \subseteq K^-(\text{Mod}(X))$ be the triangulated subcategory of complexes of flat $\mathcal{O}_X$-modules. Then by Proposition 1.2 and the lemma, part a1, L satisfies the hypotheses of [I.5.1] for the functor

$$F^\cdot \otimes^\cdot : K^-(\text{Mod}(X)) \longrightarrow K(\text{Mod}(X)),$$

and hence

$$\underline{\underline{L}}_{II}\otimes\colon\quad K^{-}(\mathrm{Mod}(X)) \times D^{-}(X) \longrightarrow D(X)$$

exists. (It is clearly functorial in $F^{\cdot}$).

By the lemma, part b1, this functor is exact in the first variable, hence passes to the quotient to give

$$\underline{\underline{L}}_{I}\underline{\underline{L}}_{II}\otimes\colon\quad D^{-}(X) \times D^{-}(X) \longrightarrow D(X).$$

Of course we can derive $\otimes$ in the first variable first, then in the second variable, and by [I.6.3] we get the same result. We will therefore use the ambiguous notation $F^{\cdot}\underline{\underline{\otimes}}G^{\cdot}$ for $\underline{\underline{L}}_{I}\underline{\underline{L}}_{II}\otimes(F^{\cdot},G^{\cdot})$.

__Definition.__ If $F^{\cdot},G^{\cdot} \in D^{-}(X)$, we define the __local hyperTor__

$$\underline{\mathrm{Tor}}_{i}(F^{\cdot},G^{\cdot}) = H^{-i}(F^{\cdot}\underline{\underline{\otimes}}G^{\cdot}) .$$

__Proposition 4.2.__ Let X be a prescheme, and let $F^{\cdot} \in \mathrm{Ob}\, K^{b}(\mathrm{Mod}(X))$. Then the following conditions are equivalent:

(i) There is a quasi-isomorphism $s\colon G^{\cdot} \to F^{\cdot}$ where $G^{\cdot}$ is a bounded complex of flat $\mathcal{O}_X$-modules.

(ii) The functor $F^{\cdot}\underline{\underline{\otimes}}^{\cdot}$ from $D^{-}(X)$ to $D^{-}(X)$ is way-out right (and hence way-out in both directions).

(iii) There is an n_o such that $\underline{\mathrm{Tor}}_{i}(F^{\cdot},G) = 0$ for all $\mathcal{O}_X$-modules G, and all $i > n_o$.

<u>Proof</u>. The proof is entirely analogous to the proof of [I.7.6] and will be left to the reader. Important points are 1) an $\mathcal{O}_X$-module F is flat if and only if $\underline{\mathrm{Tor}}_1(F,G) = 0$ for all $\mathcal{O}_X$-modules G, 2) if $Z \to B \to 0$ is a surjection of $\mathcal{O}_X$-modules with $Z \nsim B$, then there exists an $\mathcal{O}_X$-module G (for example $G = \mathcal{O}_X$) such that $Z \otimes G \nsim B \otimes G$, and 3) the analogue of [I.4.5] for flat modules fails. Instead, we use the commutative diagram

$$\begin{array}{ccc} F'^{\cdot} & \longrightarrow & F^{\cdot} \\ \downarrow & & \downarrow \\ \sigma_{\geq n}(F'^{\cdot}) & \longrightarrow & \sigma_{\geq n}(F^{\cdot}) \end{array}$$

where $F'^{\cdot}$ is a (not necessarily bounded) flat resolution of $F^{\cdot}$. Since $F^{\cdot}$ itself is a bounded complex, for n small enough, $\sigma_{\geq n}(F^{\cdot})$ is equal to $F^{\cdot}$. Hence $\sigma_{\geq n}(F'^{\cdot}) \longrightarrow F^{\cdot}$ is the required quasi-isomorphism.

<u>Remark</u>. We see from the proof that it is sufficient in (iii) to consider only quasi-coherent sheaves G. Indeed, it is sufficient to consider only the sheaves $\mathcal{O}_X$, and $k(x)$ for every point $x \in X$. Similarly, in (ii) it is sufficient to consider the restriction of the functor $F^{\cdot}\underline{\otimes}\cdot$ to $D^-_{qc}(X)$.

<u>Definition and Corollary 4.3</u>. If $F^{\cdot} \in \mathrm{Ob}\, K^b(\mathrm{Mod}(X))$ satisfies the equivalent conditions of the Proposition, we say that $F^{\cdot}$ has <u>finite Tor-dimension</u>. These complexes form a localizing subcategory of $K(\mathrm{Mod}(X))$, which we denote by $K^b(\mathrm{Mod}(X))_{fTd}$. The corresponding subcategory of $D(X)$ is denoted by $D^b(X)_{fTd}$.

<u>Proof</u>. Cf. [I.7.7].

Now let $F^{\cdot} \in \mathrm{Ob}\, K(\mathrm{Mod}(X))$, and consider the functor

$$F^{\cdot}\otimes\cdot : \quad K^b(\mathrm{Mod}(X))_{fTd} \longrightarrow K(\mathrm{Mod}(X)).$$

Let $L \subseteq K^b(\mathrm{Mod}(X))_{fTd}$ be the subcategory of complexes of flat modules. Then by Proposition 4.2 and Lemma 4.1, part a2, L satisfies the hypotheses of [I.5.1] for this functor. Hence we can take the left derived functor in the second variable, and obtain

$$\underline{\underline{L}}_{II}\otimes: \quad K(\mathrm{Mod}(X)) \times D^b(X)_{fTd} \longrightarrow D(X).$$

By Lemma 4.1, part b2, this gives rise to a functor

$$\underline{\underline{L}}_{I}\, \underline{\underline{L}}_{II}\otimes: \quad D(X) \times D^b(X)_{fTd} \longrightarrow D(X)$$

which we will also denote by $\underline{\underline{\otimes}}$.

Problem. Does the functor

$$\otimes: \ K(\mathrm{Mod}(X)) \times K^b(\mathrm{Mod}(X))_{fTd} \longrightarrow K(\mathrm{Mod}(X))$$

admit a left derived functor in the first variable? Given $G^\cdot \in \mathrm{Ob}\ K^b(\mathrm{Mod}(X))_{fTd}$, does there exist a subcategory $L \subseteq K(\mathrm{Mod}(X))$ satisfying the hypotheses of [I.5.1] for $\cdot \otimes G^\cdot$?

Proposition 4.3. Let X be a prescheme (resp. a locally noetherian prescheme), and let $F^\cdot \in D_{qc}(X)$ (resp. $D_c(X)$) and let $G^\cdot \in D^-_{qc}(X)$ (resp. $D^-_c(X)$). Assume either

a) $F^\cdot \in D^-(X)$, or

b) $G^\cdot \in D^b(X)_{fTd}$.

Then $F^\cdot \underline{\underline{\otimes}} G^\cdot \in D_{qc}(X)$ (resp. $D_c(X)$).

Proof. Using [I.7.3] as before, it is enough to show that if F,G are quasi-coherent (resp. coherent) $\mathcal{O}_X$-modules, then $\underline{\mathrm{Tor}}_i(F,G)$ is quasi-coherent (resp. coherent) for all $i \geq 0$. The question is local on X, so we may assume X affine. Then F has a resolution by quasi-coherent (resp. coherent) $\mathcal{O}_X$-modules, namely direct sums (resp. finite direct sums) of copies of $\mathcal{O}_X$. We can use this resolution to calculate the $\underline{\mathrm{Tor}}_i(F,G)$.

Suppose now that $f: X \longrightarrow Y$ is a morphism of preschemes. Then we have

$$f^*: \mathrm{Mod}(Y) \longrightarrow \mathrm{Mod}(X),$$

and we can take its left-derived functor

$$\underline{\underline{L}}f^*: D^-(Y) \longrightarrow D^-(X)$$

since there are enough flat $\mathcal{O}_Y$-modules, and they are f^*-acyclic. If f^* has finite cohomological dimension on $\mathrm{Mod}(Y)$, then we say f has <u>finite Tor-dimension</u>, and we can extend the domain of definition of $\underline{\underline{L}}f^*$ to

$$\underline{\underline{L}}f^*: D(Y) \longrightarrow D(X) .$$

<u>Proposition 4.4</u>. Let $f: X \longrightarrow Y$ be a morphism of preschemes (resp. locally noetherian preschemes). Then $\underline{\underline{L}}f^*$ takes $D^-_{qc}(Y)$ (resp. $D^-_c(Y)$) into $D^-_{qc}(X)$ (resp. $D^-_c(X)$). If f is of finite Tor-dimension, the same is true for unbounded complexes.

<u>Proof</u>. Left to reader.

§5. Relations among the derived functors.

In this section we will make a list of various natural homomorphisms and isomorphisms among the derived functors discussed in the previous sections.

Proposition 5.1. Let $f: X \longrightarrow Y$ and $g: Y \longrightarrow Z$ be two morphisms of preschemes. Then there is a natural isomorphism

$$\zeta^+ : \underline{\underline{R}}^+(g_* \cdot f_*) \xrightarrow{\sim} \underline{\underline{R}}^+ g_* \cdot \underline{\underline{R}}^+ f_*$$

of functors from $D^+(X)$ to $D^+(Z)$.

Suppose furthermore that X and Y are noetherian of finite Krull dimension. Then there is a natural isomorphism

$$\zeta : \underline{\underline{R}}(g_* \cdot f_*) \xrightarrow{\sim} \underline{\underline{R}} g_* \cdot \underline{\underline{R}} f_*$$

of functors from $D(X)$ to $D(Z)$.

Proof. We use [I.5.4b]. For the first statement, let $L \subseteq K^+(\mathrm{Mod}(X))$ be the complexes of injective sheaves, and let $M \subseteq K^+(\mathrm{Mod}(Y))$ be the complexes of g_*-acyclic sheaves. Then $f_*(L) \subseteq M$ (indeed, f_* of an injective sheaf is flasque [EGA 0_{III} 12.2.4] and a flasque sheaf is g_*-acyclic [EGA 0_{III} 12.2.1]). Then the hypotheses are satisfied, so we have ζ^+.

For the second statement, let $L \subseteq K(\mathrm{Mod}(X))$ be the complexes of sheaves which are f_*-acyclic and g_*f_*-acyclic. Let $M \subseteq K(\mathrm{Mod}(Y))$ be the complexes of g_*-acyclic sheaves. Then by the first statement of the proposition we see that $f_*(L) \subseteq M$. Indeed, if F is both f_*-acyclic and g_*f_*-acyclic, then $\underline{\underline{R}}f_*(F) \cong f_*(F)$, so for $i > 0$, $R^i g_*(f_*F) \cong R^i(g_*f_*)(F) = 0$, and f_*F is g_*-acyclic. On the other hand, one sees easily that the collection P of sheaves on X which are f_*-acyclic and g_*f_*-acyclic satisfies (i) and (ii) of [I.4.6], so L and M satisfy the hypotheses of [I.5.1], and we have the isomorphism ζ by [I.5.4b].

Proposition 5.2. Let $f\colon X \longrightarrow Y$ be a morphism of preschemes. Then there is a natural isomorphism

$$\zeta^+\colon\ \underline{\underline{R}}^+\Gamma(X,\cdot) \xrightarrow{\ \sim\ } \underline{\underline{R}}^+\Gamma(Y, \underline{\underline{R}}f_*(\cdot))$$

of functors from $D^+(X)$ to $D^+(Ab)$.

If furthermore X and Y are noetherian of finite Krull dimension, then there is a natural isomorphism

$$\zeta\colon\ \underline{\underline{R}}\Gamma(X,\cdot) \xrightarrow{\ \sim\ } \underline{\underline{R}}\Gamma(Y, \underline{\underline{R}}f_*(\cdot))$$

of functors from $D(X)$ to $D(Ab)$.

<u>Proof</u>. Similar to proof of previous proposition.

<u>Proposition 5.3</u>. Let X be a prescheme. Then there is a natural isomorphism

$$\underline{\underline{R}}\,\mathrm{Hom}^{\cdot}(F^{\cdot},G^{\cdot}) \xrightarrow{\sim} \underline{\underline{R}}\Gamma(X, \underline{\underline{R}}\,\underline{\mathrm{Hom}}^{\cdot}(F^{\cdot},G^{\cdot}))$$

of bi-functors from $D^{-}(X)^{\circ} \times D^{+}(X)$ to $D(Ab)$.

<u>Proof</u>. Use [I.5.4]. Note that for any two sheaves F and G, $\mathrm{Hom}(F,G) = \Gamma(\underline{\mathrm{Hom}}(F,G))$. Also note that if G is injective, then $\underline{\mathrm{Hom}}(F,G)$ is flasque [G, II 7.3.2], and hence Γ-acyclic.

<u>Proposition 5.4</u>. Let $f: X \longrightarrow Y$ and $g: Y \longrightarrow Z$ be morphisms of preschemes. Then there is a natural isomorphism

$$\zeta^{-}: \underline{\underline{L}}^{-}(f^{*}\circ g^{*}) \xrightarrow{\sim} \underline{\underline{L}}^{-}f^{*} \circ \underline{\underline{L}}^{-}g^{*}$$

of functors from $D^{-}(Z)$ to $D^{-}(X)$.

If furthermore f and g have finite Tor-dimension (see §4) then so does $g\circ f$, and there is a natural isomorphism

$$\zeta: \underline{\underline{L}}(g\circ f)^{*} \xrightarrow{\sim} \underline{\underline{L}}f^{*} \circ \underline{\underline{L}}g^{*}$$

of functors from $D^{-}(Z)$ to $D^{-}(X)$.

<u>Proof</u>. Left to reader. (Note that g^{*} takes flat $\mathcal{O}_{Z}$-modules into flat $\mathcal{O}_{Y}$-modules.)

Proposition 5.5. Let $f: X \longrightarrow Y$ be a morphism of preschemes, with X noetherian of finite Krull dimension. Then there is a natural functorial homomorphism

$$\underline{\underline{R}}f_* \, \underline{\underline{R}} \, \underline{\text{Hom}}^{\cdot}_X(F^{\cdot},G^{\cdot}) \longrightarrow \underline{\underline{R}} \, \underline{\text{Hom}}^{\cdot}_Y(\underline{\underline{R}}f_*F^{\cdot}, \underline{\underline{R}}f_*G^{\cdot})$$

for $F^{\cdot} \in D^-(X)$ and $G^{\cdot} \in D^+(X)$.

Proof. Note first that our hypotheses on $f,Y,F^{\cdot}$, and $G^{\cdot}$ ensure that both objects above are defined (cf. §2 and §3 above). We wish to define a morphism between two functors from $D^-(X)^{\circ} \times D^+(X)$ to $D(Y)$. Let $L \subseteq K^-(\text{Mod}(X))$ be the complexes of f_*-acyclic objects, and let $M \subseteq K^+(\text{Mod}(X))$ be the complexes of injective objects. Then, as we have seen before ([I, §5] and §2 above) the natural functors

$$L_{qis} \longrightarrow D^-(X)$$

and $$M_{Qis} \longrightarrow D^+(X)$$

are equivalences of categories. Hence it will be sufficient to define a morphism between the extensions of our functors to functors from $L_{Qis}^{\circ} \times M_{Qis}$ to $D(Y)$. But now, using [I.3.4] applied to the triangulated categories L and M, we see that it is enough to define a morphism of functors

$$\underline{\underline{R}}f_* \, \underline{\underline{R}} \, \underline{\mathrm{Hom}}^{\cdot}_X(QF^{\cdot}, QG^{\cdot}) \longrightarrow \underline{\underline{R}} \, \underline{\mathrm{Hom}}^{\cdot}_Y(\underline{\underline{R}}f_*QF^{\cdot}, \underline{\underline{R}}f_*QG^{\cdot}) ,$$

(where Q denotes the localization functor from L to L_{Qis} or M to M_{Qis}) for $F^{\cdot} \in L$ and $G^{\cdot} \in M$.

We now make explicit the morphisms ξ between a functor and its derived functor (cf. definition of derived functor, [I.5]) and obtain the following diagram:

$$\begin{array}{ccc}
Qf_*\underline{\mathrm{Hom}}^{\cdot}(F^{\cdot},G^{\cdot}) & \xrightarrow[(1)]{\xi_{f_*}} & \underline{\underline{R}}f_* \, Q \, \underline{\mathrm{Hom}}^{\cdot}(F^{\cdot},G^{\cdot}) \\
\Big\downarrow \varphi & & (2) \Big\downarrow \xi_{\underline{\mathrm{Hom}}^{\cdot}} \\
Q \, \underline{\mathrm{Hom}}^{\cdot}(f_*F^{\cdot},f_*G^{\cdot}) & & \underline{\underline{R}}f_* \, \underline{\underline{R}} \, \underline{\mathrm{Hom}}^{\cdot}(QF^{\cdot},QG^{\cdot}) \\
(3) \Big\downarrow \xi_{\underline{\mathrm{Hom}}^{\cdot}} & & \vdots \\
\underline{\underline{R}} \, \underline{\mathrm{Hom}}^{\cdot}(Qf_*F^{\cdot},Qf_*G^{\cdot}) & & \vdots \\
(4) \Big\downarrow \xi^{II}_{f_*} & & \Big\downarrow \\
\underline{\underline{R}} \, \underline{\mathrm{Hom}}^{\cdot}(Qf_*F^{\cdot},\underline{\underline{R}}f_*QG^{\cdot}) & \xrightarrow[(5)]{\xi^{I}_{f_*}} & \underline{\underline{R}} \, \underline{\mathrm{Hom}}^{\cdot}(\underline{\underline{R}}f_*QF^{\cdot},\underline{\underline{R}}f_*QG^{\cdot}) .
\end{array}$$

Here φ is deduced from the well-known natural map

$$f_* \, \underline{\mathrm{Hom}}_X(F,G) \longrightarrow \underline{\mathrm{Hom}}_Y(f_*F,f_*G)$$

for any two sheaves of $\mathcal{O}_X$-modules F,G. Now we use the last statement of [I.5.1] to deduce that certain of the ξ's are isomorphisms.

(1) Since $G^{\cdot}$ is made of injective sheaves, $\underline{Hom}^{\cdot}(F^{\cdot},G^{\cdot})$ is made of flasque sheaves [G, II.7.3.2] (note only finite products are involved since $F^{\cdot} \in K^{-}$ and $G^{\cdot} \in K^{+}$). But flasque sheaves are f_*-acyclic, so they can be used to calculate $\underline{\underline{R}}f_*$, and hence the map ξ_{f_*} denoted by (1) above is an isomorphism.

(2) $G^{\cdot}$ is injective (or more precisely "M satisfies the hypotheses of [I.5.1] for $\underline{Hom}^{\cdot}(F^{\cdot},\cdot)$") and so $\xi_{\underline{Hom}^{\cdot}}$ is an isomorphism.

(3) We can say nothing here.

(4) $G^{\cdot}$ is injective, so ξ_{f_*} is an isomorphism.

(5) $F^{\cdot}$ is made of f_*-acyclic sheaves, so ξ_{f_*} here is an isomorphism.

Therefore, since (1), (2), (4), and (5) are isomorphisms, there is a unique

$$\psi\colon \underline{\underline{R}}f_* \, \underline{\underline{R}} \, \underline{Hom}^{\cdot}(QF^{\cdot},QG^{\cdot}) \longrightarrow \underline{\underline{R}} \, \underline{Hom}^{\cdot}(\underline{\underline{R}}f_*QF^{\cdot}, \underline{\underline{R}}f_*QG^{\cdot})$$

making the diagram commutative. This is the desired morphism of functors.

<u>Remark</u>. We have given the above proof in some detail to show the method. Faced with a similar situation in the sequel, we will say simply "we may assume $F^{\cdot}$ is made of f_*-acyclic objects, and

$G^\cdot$ on injective objects", and then we will drop all Q's, and write "=" for any ξ which is an isomorphism. In other words, we use the convention that we may erase the $\underline{\underline{R}}$ before a functor applied to an argument for which ξ is an isomorphism. So if $F^\cdot$ is a complex of f_*-acyclic objects, we will write $\underline{\underline{R}}f_*(F^\cdot) = f_*F^\cdot$.

Proposition 5.6 (Projection formula). Let $f: X \longrightarrow Y$ be a quasi-compact morphism of noetherian preschemes, of finite Krull dimension. Then there is a natural functorial isomorphism

$$\underline{\underline{R}}f_*(F^\cdot) \underline{\underline{\otimes}}_Y G^\cdot \xrightarrow{\sim} \underline{\underline{R}}f_*(F^\cdot \underline{\underline{\otimes}}_X \underline{\underline{L}}f^*G^\cdot)$$

for $F^\cdot \in D^-(X)$ and $G^\cdot \in D^-_{qc}(Y)$.

Proof. Note first that both sides are defined. To define the morphism, we may assume that $F^\cdot$ is a complex of f_*-acyclic sheaves, and that $G^\cdot$ is a complex of $\mathcal{O}_Y$-flat sheaves. Then we get the morphism by composing the usual projection formula for sheaves with a suitable ξ:

$$f_*(F^\cdot) \otimes_Y G^\cdot \longrightarrow f_*(F^\cdot \otimes_X f^*G^\cdot) \xrightarrow{\xi_{f_*}} \underline{\underline{R}}f_*(F^\cdot \otimes_X f^*G^\cdot)$$

using the conventions of the remark above.

To show it is an isomorphism, we must show for each i that the map of sheaves

$$H^i(\underline{\underline{R}}f_*(F^\cdot)\underline{\underline{\otimes}}_Y G^\cdot) \longrightarrow H^i(\underline{\underline{R}}f_*(F^\cdot \underline{\underline{\otimes}}_X \underline{\underline{L}}f^* G^\cdot))$$

is an isomorphism. But this question is local on Y. Furthermore, the functors involved are all compatible with localization on Y, so we may assume Y affine. Now, going back to the original morphism, note that both sides are way-out left functors in $G^\cdot$. Note also that every quasi-coherent $\mathcal{O}_Y$-module is a quotient of a free $\mathcal{O}_Y$-module, since Y is affine. Hence, using the Lemma on Way-Out Functors [I.7.1 (ii) and (iv), dual statement], we reduce to the case where $G^\cdot$ is reduced to a single sheaf G, which is a free $\mathcal{O}_Y$-module. But now X, Y, and f are quasi-compact, so everything commutes with infinite direct sums, and we reduce to the case $G = \mathcal{O}_Y$. (The noetherian hypothesis ensures that $R^i f_*$ commutes with direct sums, cf. [T.3.6.2] and [G,II §4.12]). Then we have $\underline{\underline{R}}f_*(F^\cdot)$ on each side and we are done.

<u>Remark</u>. Again we have given a proof in some detail to show the method. In the sequel, we may leave many of these details to the reader.

Corollary 5.7. Let f, X, Y be as in the Proposition, and assume that f has finite Tor-dimension. Let $F^{\cdot} \in D^b(X)$ have finite Tor-dimension (see Definition 4.3 above). Then $\underline{\underline{R}}f_*F^{\cdot} \in D^b(Y)$ also has finite Tor-dimension.

Proof. We use condition (ii) of Proposition 4.2. Since f has finite Tor-dimension, the functor on $D^-_{qc}(Y)$

$$G^{\cdot} \longmapsto \underline{\underline{L}}f^*G^{\cdot}$$

is way-out right. Since $F^{\cdot}$ also has finite Tor-dimension,

$$G^{\cdot} \longmapsto F^{\cdot} \underline{\underline{\otimes}}_X \underline{\underline{L}}f^*G^{\cdot}$$

is way-out right, and so

$$G^{\cdot} \longmapsto \underline{\underline{R}}f_*(F^{\cdot} \underline{\underline{\otimes}}_X \underline{\underline{L}}f^*G^{\cdot})$$

is way-out right. By the Proposition, this implies that the functor

$$G^{\cdot} \longmapsto \underline{\underline{R}}f_*(F^{\cdot}) \underline{\underline{\otimes}} G^{\cdot}$$

is way-out right (for $G^{\cdot} \in D^-_{qc}(Y)$). But this implies that $\underline{\underline{R}}f_*(F^{\cdot})$ has finite Tor-dimension (using the Remark following Proposition 4.2).

Problem. Let f, X, Y be as in the Proposition, and assume that f has finite Tor-dimension. Then the functors $\underline{\underline{R}}f_*(F^\cdot)\underline{\underline{\otimes}}_Y G^\cdot$ and $\underline{\underline{R}}f_*(F^\cdot \underline{\underline{\otimes}}\, \underline{\underline{L}}f^*G^\cdot)$ are both defined for $F^\cdot \in D^b(X)_{fTd}$, and $G^\cdot \in D_{qc}(Y)$, using the result of the Corollary. Are they isomorphic? The problem is to define a morphism between them, since once we have a morphism, it is easy to prove that it is an isomorphism, using the lemma on way-out functors.

Proposition 5.8. Let $f: X \longrightarrow Y$ be a flat morphism of preschemes. Then there is a natural functorial homomorphism

$$f^* \underline{\underline{R}}\ \underline{\mathrm{Hom}}^\cdot_Y(F^\cdot, G^\cdot) \longrightarrow \underline{\underline{R}}\ \underline{\mathrm{Hom}}^\cdot_X(f^*F^\cdot, f^*G^\cdot)$$

for $F^\cdot \in D(Y)$ and $G^\cdot \in D^+(Y)$. If furthermore Y is locally noetherian, and $F^\cdot \in D_c^-(Y)$, it is an isomorphism. (We write f^* instead of $\underline{\underline{L}}f^*$ since it is an exact functor.)

Proof. To define the map, we may assume $G^\cdot$ is a complex of injective sheaves, and then use the natural map of sheaves

$$f^*\underline{\mathrm{Hom}}_Y(F,G) \longrightarrow \underline{\mathrm{Hom}}_X(f^*F, f^*G).$$

To show the isomorphism, we may assume that Y is affine, and then we reduce to the case $F = \mathcal{O}_Y$ by the lemma on way-out functors.

<u>Proposition 5.9</u>. Let $f: X \longrightarrow Y$ be a morphism of preschemes. Then there is a natural functorial isomorphism

$$\underline{\underline{L}}f^*(F^\cdot) \underline{\underline{\otimes}}_X \underline{\underline{L}}f^*(G^\cdot) \xrightarrow{\sim} \underline{\underline{L}}f^*(F^\cdot \underline{\underline{\otimes}}_Y G^\cdot)$$

for $F^\cdot, G^\cdot \in D^-(Y)$.

<u>Proof</u>. Left to reader.

<u>Proposition 5.10</u>. Let $f: X \longrightarrow Y$ be a morphism of noetherian preschemes of finite Krull dimension. Then there is a natural functorial homomorphism

$$\rho: F^\cdot \longrightarrow \underline{\underline{R}}f_* \underline{\underline{L}}f^* F^\cdot$$

for $F^\cdot \in D^-(Y)$, which gives rise by Proposition 5.5 to a natural functorial homomorphism

$$\tau: \underline{\underline{R}}f_* \underline{\underline{R}} \underline{\mathrm{Hom}}^\cdot_X(\underline{\underline{L}}f^* F^\cdot, G^\cdot) \longrightarrow \underline{\underline{R}} \underline{\mathrm{Hom}}^\cdot_Y(F^\cdot, \underline{\underline{R}}f_* G^\cdot)$$

for $F^\cdot \in D^-(Y)$ and $G^\cdot \in D^+(X)$.

If furthermore $F^\cdot \in D^-_c(Y)$, then τ is an isomorphism.

<u>Proof</u>. To define ρ, we may assume $F^\cdot$ is a complex of flat $\mathcal{O}_Y$-modules, compose the natural map $F^\cdot \longrightarrow f_*f^*F^\cdot$ with ξ_{f_*}. To check that τ is an isomorphism, we may assume Y affine, and then reduce to the case $F^\cdot = \mathcal{O}_Y$. Then $\underline{\underline{L}}f^*F^\cdot = \mathcal{O}_X$, and we have simply $\underline{\underline{R}}f_*G^\cdot$ on each side.

<u>Corollary 5.11</u>. Under the hypotheses of the Proposition, we have

$$\mathrm{Hom}_{D(X)}(\underline{\underline{L}}f^*F^\cdot, G^\cdot) \xrightarrow{\sim} \mathrm{Hom}_{D(Y)}(F^\cdot, \underline{\underline{R}}f_*G^\cdot) ,$$

in other words, $\underline{\underline{L}}f^*$ and $\underline{\underline{R}}f_*$ are adjoint functors from $D_c^-(Y)$ to $D_c^-(X)$ and $D^+(X)$ to $D^+(Y)$, respectively.

<u>Proof</u>. Apply $H^o\underline{\underline{R}}\Gamma$ to both sides of the isomorphism τ of the Proposition, and use Propositions 5.2, 5.3, and [I.6.4].

<u>Proposition 5.12</u>. Let $f\colon X \longrightarrow Y$ be a morphism of finite type of noetherian preschemes of finite Krull dimension. Let $u\colon Y' \longrightarrow Y$ be a flat morphism, let $X' = X \times_Y Y'$, and let v, g be the projections, as shown. Then there is a natural functorial isomorphism

$$\begin{array}{ccc} X' & \xrightarrow{v} & X \\ \downarrow{\scriptstyle g} & & \downarrow{\scriptstyle f} \\ Y' & \xrightarrow{u} & Y \end{array}$$

$$u^* \underline{\underline{R}}f_*F^\cdot \xrightarrow{\sim} \underline{\underline{R}}g_*v^*F^\cdot$$

for $F^\cdot \in D_{qc}(X)$.

Proof. To define the morphism, we may assume $F^{\cdot}$ is a complex of f_*-acyclic sheaves, and use the natural map $u^*f_*F^{\cdot} \longrightarrow g_*v^*F^{\cdot}$ followed by ξ_{g_*}. Both sides are way-out in both directions, so to prove the isomorphism we reduce by the lemma on way-out functors to the case of a single quasi-coherent sheaf F on Y. Then we must show that for each i,

$$H^i(u^*\underline{\underline{R}}f_*F) \xrightarrow{\sim} H^i(\underline{\underline{R}}g_*v^*F) .$$

This is [EGA III 1.4.15]. (Recall that the proof uses Čech cohomology, hence the quasi-coherence hypothesis.)

Proposition 5.13: Let X be a prescheme. Then there are natural functorial isomorphisms

$$F^{\cdot} \underline{\underline{\otimes}} G^{\cdot} \xrightarrow{\sim} G^{\cdot} \underline{\underline{\otimes}} F^{\cdot}$$

and $$F^{\cdot} \underline{\underline{\otimes}} (G^{\cdot} \underline{\underline{\otimes}} H^{\cdot}) \xrightarrow{\sim} (F^{\cdot} \underline{\underline{\otimes}} G^{\cdot}) \underline{\underline{\otimes}} H^{\cdot}$$

for all $F^{\cdot}, G^{\cdot}$, and $H^{\cdot} \in D^-(X)$.

Proof. Left to reader.

Proposition 5.14. Let X be a prescheme. Then there is a natural functorial homomorphism

$$\underline{\underline{R}}\ \underline{\mathrm{Hom}}^{\cdot}(F^{\cdot}, G^{\cdot}) \underline{\underline{\otimes}} H^{\cdot} \longrightarrow \underline{\underline{R}}\ \underline{\mathrm{Hom}}^{\cdot}(F^{\cdot}, G^{\cdot} \underline{\underline{\otimes}} H^{\cdot})$$

for $F^\cdot \in D(X)$, $G^\cdot \in D^+(X)$, and $H^\cdot \in D^b(X)_{fTd}$. If furthermore X is locally noetherian, and $F^\cdot \in D_c^-(X)$, then it is an isomorphism.

Proof. Left to reader.

Proposition 5.15. Let X be a locally noetherian prescheme, and assume that every coherent sheaf on X is a quotient of a locally free sheaf of finite rank (lffr). Then there is a natural functorial isomorphism

$$\underline{\underline{R}}\ \underline{Hom}^\cdot(F^\cdot, \underline{\underline{R}}\ \underline{Hom}^\cdot(G^\cdot,H^\cdot)) \xrightarrow{\sim} \underline{\underline{R}}\ \underline{Hom}^\cdot(F^\cdot \underline{\underline{\otimes}} G^\cdot, H^\cdot)$$

for $F^\cdot$ and $G^\cdot \in D_c^-(X)$, and $H^\cdot \in D^+(X)$.

Proof. To define the morphism, we use the result of the exercise in §3 above. We take resolutions of $F^\cdot$ and $G^\cdot$ by lffr's, and use them to calculate $\underline{\underline{R}}\ \underline{Hom}^\cdot$. Note that any lffr is flat, and so if $F^\cdot, G^\cdot$ are complexes of lffr's, then $F^\cdot \underline{\underline{\otimes}} G^\cdot = F^\cdot \otimes G^\cdot$, which is also a complex of lffr's. For the isomorphism, we use the lemma on way-out functors and reduce to the case $F^\cdot = \mathcal{O}_X$.

Proposition 5.16. Let X be a prescheme, and let $L^\cdot$ be a bounded complex of locally free sheaves of finite rank. Let $L^{\cdot\vee} = \underline{Hom}^\cdot(L^\cdot, \mathcal{O}_X)$. Then there are natural functorial isomorphisms

$$\underline{\underline{R}}\ \underline{Hom}^{\cdot}(F^{\cdot},G^{\cdot}) \otimes L^{\cdot} \xrightarrow{\sim} \underline{\underline{R}}\ \underline{Hom}^{\cdot}(F^{\cdot},G^{\cdot}\otimes L^{\cdot})$$

$$\xrightarrow{\sim} \underline{\underline{R}}\ \underline{Hom}^{\cdot}(F^{\cdot}\otimes L^{\cdot\vee}, G^{\cdot})$$

for all $F^{\cdot} \in D^{-}(X)$ and $G^{\cdot} \in D^{+}(X)$.

Proof. Easy once one notes that the corresponding formulae hold for sheaves, and if G is an injective sheaf, and L a lffr, then $G\otimes L$ is injective.

§6. Compatibilities among the relations of §5.

In situations involving three or more derived functors, there may be different ways of composing the homomorphisms and isomorphisms of section 5 to obtain a homomorphism or isomorphism of functors. One would like to know that the result is independent of any choices. We give three examples.

1. Let $f: X \longrightarrow Y$, $g: Y \longrightarrow Z$, and $h: Z \longrightarrow W$ be three morphisms of preschemes. Then by Proposition 5.1 there are isomorphisms

$$\begin{array}{ccc} \underline{\underline{R}}^+(h_* g_* f_*) & \xrightarrow{\sim} & \underline{\underline{R}}^+ h_* \, \underline{\underline{R}}^+(g_* f_*) \\ \downarrow & & \downarrow \wr \\ \underline{\underline{R}}^+(h_* g_*) \underline{\underline{R}}^+(f_*) & \xrightarrow{\sim} & \underline{\underline{R}}^+ h_* \, \underline{\underline{R}}^+ g_* \, \underline{\underline{R}}^+ f_* \end{array} .$$

We would like to know that this diagram is commutative.

2. Let $f: X \rightarrow Y$ and $g: Y \rightarrow Z$ be morphisms of noetherian preschemes of finite Krull dimension. Then there are functorial homomorphisms

$$
\begin{array}{ccc}
\underline{\underline{R}}(g_*f_*)\underline{\underline{R}}\ \underline{\mathrm{Hom}}^{\bullet}_X(F^{\bullet},G^{\bullet}) & \xrightarrow{\ \sim\ } & \underline{\underline{R}}g_*\ \underline{\underline{R}}f_*\ \underline{\underline{R}}\ \underline{\mathrm{Hom}}^{\bullet}_X(F^{\bullet},G^{\bullet}) \\
\Big\downarrow & & \Big\downarrow \\
 & & \underline{\underline{R}}g_*\ \underline{\underline{R}}\ \underline{\mathrm{Hom}}^{\bullet}_Y(\underline{\underline{R}}f_*F^{\bullet},\underline{\underline{R}}f_*G^{\bullet}) \\
 & & \Big\downarrow \\
\underline{\underline{R}}\ \underline{\mathrm{Hom}}^{\bullet}_Z(\underline{\underline{R}}(g_*f_*)F^{\bullet},\underline{\underline{R}}(g_*f_*)G^{\bullet}) & \xrightarrow{\ \sim\ } & \underline{\underline{R}}\ \underline{\mathrm{Hom}}^{\bullet}_Z(\underline{\underline{R}}g_*\underline{\underline{R}}f_*F^{\bullet},\underline{\underline{R}}g_*\ \underline{\underline{R}}f_*G^{\bullet})
\end{array}
$$

for $F^{\bullet} \in D^-(X)$ and $G^{\bullet} \in D^+(X)$. Here the horizontal arrows are deduced from Proposition 5.1, and the vertical arrows from Proposition 5.5, and we have tacitly included in the left-hand vertical arrow a double use of the natural isomorphism $g_*f_* \cong (gf)_*$. We would like to know that this diagram is commutative.

3. Let X be a prescheme. Then there are functorial isomorphisms

$$
\begin{array}{ccc}
F^{\bullet}\ \underline{\underline{\otimes}}(G^{\bullet}\underline{\underline{\otimes}}(H^{\bullet}\underline{\underline{\otimes}}I^{\bullet})) & \xrightarrow{\ \sim\ } & F^{\bullet}\ \underline{\underline{\otimes}}((G^{\bullet}\underline{\underline{\otimes}}H^{\bullet})\underline{\underline{\otimes}}I^{\bullet}) \\
\Big\downarrow & & \Big\downarrow\wr \\
 & & (F^{\bullet}\ \underline{\underline{\otimes}}\ (G^{\bullet}\underline{\underline{\otimes}}H^{\bullet}))\ \underline{\underline{\otimes}}\ I^{\bullet} \\
 & & \Big\downarrow\wr \\
(F^{\bullet}\underline{\underline{\otimes}}G^{\bullet})\ \underline{\underline{\otimes}}\ (H^{\bullet}\underline{\underline{\otimes}}I^{\bullet}) & \xrightarrow{\ \sim\ } & ((F^{\bullet}\underline{\underline{\otimes}}G^{\bullet})\underline{\underline{\otimes}}H^{\bullet})\ \underline{\underline{\otimes}}\ I^{\bullet}
\end{array}
$$

for $F^{\bullet},G^{\bullet},H^{\bullet}$, and $I^{\bullet} \in D^-(X)$, by Proposition 5.13. We would like to know that this diagram is commutative.

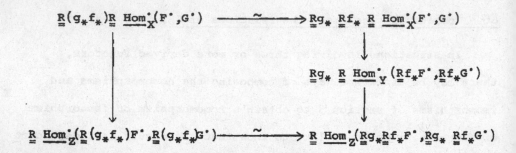

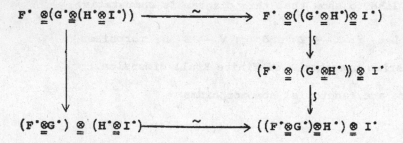

In the first example, the commutativity follows from the Remark after [I.5.4]. In the second and third examples, we note that the homomorphisms and isomorphisms referred to are uniquely determined by the corresponding natural maps for sheaves, and by the condition that they commute with the relevant morphisms ξ which occur in the definition of the derived functor. The required commutativities then follow from the analogous results for sheaves, which we may assume known.

Now these examples are only three of many more similar compatibilities which will come immediately to the reader's mind. I could make a big list, and in principle could prove each one on the list. However, I would be sure to need some more later, and already the list of ones I can think of offhand is too cumbersome to write down. And since the chore of inventing these diagrams and checking their commutativity is almost mechanical, the reader would not want to read them, nor I write them. It would be comforting to know that such a list existed, or to have a meta-theorem saying that any such diagram one would dream up is commutative. However, both of these possibilities seem of an order of complexity too great to treat in these notes.

Unfortunately, I will have to use many of these compatibilities in an essential way in what is to follow. Perhaps for each theorem in the sequel one could make a list of exactly which compatibilities are needed, and verify them, but even that is too clumsy at this stage. So I must ask the reader's indulgence. I believe in the truth of the theorems stated, and I hope to convince him of their truth also. But I have not verified every commutative diagram which is necessary for a rigorous proof, and I do not suppose that any reader will have the patience to do so either.

In particular, I believe that all reasonable compatibilities one could imagine between the "natural" homomorphisms and isomorphisms of section 5 are true, and so in the sequel I will write simply "=" instead of naming the isomorphism (provided, of course, the relevant hypotheses are satisfied). However, in later chapters we will deal with more homomorphisms and isomorphisms, where the compatibilities one can imagine are not always obvious, and may even be false (e.g., where there is a question of sign). Hence I will name those isomorphisms and list (in principle) the compatibilities we will need, and keep track of them.

For a completely satisfactory treatment of this question of "natural" isomorphisms and their compatibilities, we must await future developments. Mac Lane [12 pp. 14,15] refers to the problem in the context of the associativity of the tensor product, where he says that he does not know even a finite number of diagrams such as the one in example 3 above, which together imply that all such diagrams involving associativities are commutative. Perhaps the language of fibred categories [SGA 60-61, exposé VI] or the techniques of [Giraud, thesis] will supply what is needed.

§7. Injective sheaves on a locally noetherian prescheme.

In this section we give the structure of the injective objects in the category Mod(X) of all $\mathcal{O}_X$-modules on a locally noetherian prescheme X. We show in particular that every quasi-coherent $\mathcal{O}_X$-module can be embedded in a quasi-coherent injective $\mathcal{O}_X$-module, and hence that the natural functor

$$D^+(\mathrm{Qco}(X)) \longrightarrow D^+_{qc}(X)$$

is an equivalence of categories.

We use results of Gabriel [5], which in turn were inspired from results of Matlis [13] in the case of noetherian rings.

<u>Definition</u>. Let A be an abelian category, and let M be an object of A. An <u>injective hull</u> of M is an injective object I of A, together with a monomorphism $M \rightarrow I$, such that if N is a non-zero subobject of I, then $M \cap N \neq 0$.

<u>Theorem 7.1</u> [5 , Ch. II, Thm. 2]. If A is an abelian category with generators and exact direct limits, then every object has an injective hull.

<u>Examples</u>. The category of modules over a commutative ring has generators and exact direct limits, hence has injective hulls.

The category Mod(X) of $\mathcal{O}_X$-modules on an arbitrary prescheme X has generators and exact direct limits, hence has injective hulls.

Definition [5, Ch. II, §4]. An object of an abelian category A is noetherian if every ascending chain of subobjects is stationary. An abelian category A is locally noetherian if it has exact direct limits, and has a family of generators consisting of noetherian objects of A.

Examples. The category of modules over a noetherian ring is locally noetherian — the ring itself is a noetherian generator for the category.

The category Qco(X) of quasi-coherent sheaves on a noetherian prescheme X is locally noetherian, with the coherent sheaves forming a family of noetherian generators. This example was studied by Gabriel [5, Ch. VI].

Proposition 7.2 [5, Ch. IV, Prop. 6]. Any direct sum of injective objects in a locally noetherian category is injective.

Theorem 7.3 [5, Ch. IV, Thm. 2]. Let A be a locally noetherian category. Then every injective object I of A is isomorphic to a direct sum $\bigoplus_{j \in J} I_j$ of indecomposable injectives

I_j. Furthermore, if $\bigoplus_{k \in K} I_k$ is a second such decomposition, then there is a bijection $h: J \to K$ such that $I_j \cong I_{h(j)}$ for each j.

Proposition 7.4 [13]. Let A be a noetherian ring. Then the indecomposable injective A-modules are precisely the injective hulls $I(\mathfrak{p})$ of $k(\mathfrak{p})$ over A, where $\mathfrak{p}$ ranges over the prime ideals of A.

We also have information about the structure of one of these injective hulls [Matlis, ibid.]:

Proposition 7.5. Let A be a noetherian ring, let $\mathfrak{p}$ be a prime ideal of A, and let I be an injective hull of the residue field $k(\mathfrak{p})$ of $\mathfrak{p}$. Then one can write I as a direct limit of the submodules

$$0 \subseteq E^1 \subseteq E^2 \subseteq \cdots$$

where $E^i = \mathrm{Hom}_A(A_{\mathfrak{p}}/\mathfrak{p}^i A_{\mathfrak{p}}, I)$ is an Artin module over the local ring $A_{\mathfrak{p}}$, and where, for each i,

$$\dim_{k(\mathfrak{p})}(E^{i+1}/E^i) = \dim_{k(\mathfrak{p})}((\mathfrak{p}^i/\mathfrak{p}^{i+1}) \otimes_A k(\mathfrak{p})).$$

Now we apply these results to locally noetherian preschemes, giving first a special case.

Definition. Let X be a prescheme, x a point of X, and M and $\mathcal{O}_{x,X}$-module. Then we define $\underline{i_x(M)}$ to be the sheaf $i_*(\tilde{M})$ on X, where i: Spec $\mathcal{O}_{x,X} \longrightarrow X$ is the natural inclusion, and $\tilde{M}$ is the sheaf on Spec $\mathcal{O}_{x,X}$ associated to M.

Remark. We will be particularly interested in the case where Supp M is just the closed point x of Spec $\mathcal{O}_{x,X}$. In that case $\Gamma(U,i_x(M)) = M$ if $x \in U$, and 0 otherwise, i.e., it is a simple sheaf on the closed subset $\{x\}^-$ of X. Moreover, for any $\mathcal{O}_X$-module F, we have

$$\mathrm{Hom}_{\mathcal{O}_X}(F, i_x(M)) = \mathrm{Hom}_{\mathrm{Spec}\, \mathcal{O}_{x,X}}(i^*(F),\tilde{M}) = \mathrm{Hom}_{\mathcal{O}_{x,X}}(F_x,M)$$

where F_x is the stalk of F at x.

Proposition 7.6. Let X be a locally noetherian prescheme, let x be a point of X, and let I be an injective hull of $k(x)$ over the local ring $\mathcal{O}_{x,X}$ of x. Then $i_x(I)$ is a (quasi-coherent) injective $\mathcal{O}_X$-module.

<u>Proof</u>. By Proposition 7.5 above, I has support at the closed point of Spec $\mathcal{O}_{x,X}$. Therefore, if $F \subseteq G$ are two $\mathcal{O}_X$-modules, and $\varphi: F \longrightarrow i_x(I)$ is a homomorphism of sheaves, φ gives a map $F_x \longrightarrow I$, which extends to a map $G_x \longrightarrow I$ since I is an injective $\mathcal{O}_{x,X}$-module. But by the Remark above, this gives a map of G to $i_x(I)$ extending φ, so $i_x(I)$ is an injective $\mathcal{O}_X$-module.

<u>Lemma 7.7</u>. Let X be a noetherian prescheme, and let $G \subseteq \mathcal{O}_X$ be a sheaf of ideals (not necessarily quasi-coherent). Then there are a finite number of open subsets $U_i \subseteq X$, and a finite number of sections $s_{ij} \in \Gamma(U_i,G)$ for each i, such that at each point $x \in X$, the sections s_{ij}, for those i such that $x \in U_i$, generate the stalk G_x as an $\mathcal{O}_{x,X}$-module.

<u>Proof</u>. Since X can be covered by a finite number of open affines, we may as well assume X is affine, equal to Spec A for a suitable noetherian ring A.

Let $x \in X$. For each open affine neighborhood U of x of the form $X_f = \text{Spec } A_f$ for $f \in A$, consider the ideal $\mathfrak{a}(U) = \rho^{-1}\Gamma(U,G)$ of A, where

$$\rho: A = \Gamma(X,\mathcal{O}_X) \longrightarrow \Gamma(U,\mathcal{O}_X)$$

is the natural restriction. Clearly a smaller neighborhood gives a larger ideal. Hence by the a.c.c. in A, there is such a neighborhood U_x of x giving a maximal $\mathfrak{a}(U_x) = \mathfrak{a}_x$. Hence, since the rings $\Gamma(U,\mathcal{O}_X)$ are localizations of A, we have for every such open neighborhood $U \subseteq U_x$ of x that

$$(1) \qquad \Gamma(U,G) = \mathfrak{a}_x \otimes_A \Gamma(U,\mathcal{O}_X) = \Gamma(U_x,G)\otimes_{\Gamma(U_x,\mathcal{O}_X)}\Gamma(U,\mathcal{O}_X).$$

Now if x' is a specialization of x in U_x, i.e., $x' \in \{x\}^- \cap U_x$, then every neighborhood of x' contains x. So by (1) we see that the stalk $G_{x'}$ can be generated by sections $s_i \in \Gamma(U_x,G)$, which, being an ideal in a noetherian ring, has a finite number of generators. (Recall that open sets of the form X_f form a base for the topology of X.) Thus we have established the following result:

(*) For each point $x \in X$, one can find an open affine neighborhood U_x of x, and a finite number of sections $s_i \in \Gamma(U_x, G)$, such that for every $x' \in \{x\}^- \cap U_x$, the s_i generate the stalk $G_{x'}$.

Now we prove the statement of the Lemma. By quasi-compacity, we can find an open subset U of X (possibly empty) which is maximal with the property that the lemma is true for $G|_U$. Suppose that $U \neq X$. Let x be the generic point of an irreducible component of $X-U$, and choose a neighborhood U_x with the property (*) above, and also so small that it does not meet any other irreducible component of $X-U$. Then every point of $U \cup U_x$ lies either in U or in $\{x\}^- \cap U_x$, so that the lemma is also true for $G|_{U \cup U_x}$. ✱ We conclude that $U = X$, which proves the Lemma.

<u>Theorem 7.8</u>. Let X be a locally noetherian prescheme. Then the category $\mathrm{Mod}(X)$ of all $\mathcal{O}_X$-modules is a locally noetherian category.

<u>Proof</u>. $\mathrm{Mod}(X)$ clearly has exact direct limits. We must provide it with a family of noetherian generators. I claim the sheaves $\mathcal{O}_U$, where U is a noetherian open affine of X, and $\mathcal{O}_U$ is the sheaf $\mathcal{O}_X$ on U and zero outside, will do. They clearly form a family of generators. To see that $\mathcal{O}_U$ is noetherian, let $G_1 \subseteq G_2 \subseteq \cdots$ be an increasing sequence of subsheaves. We may assume $X = U$ is noetherian. Let $G = \bigcup G_k$. Then by the Lemma, there are a finite number of open sets U_i and sections $s_{ij} \in \Gamma(U_i, G)$ which generate the stalk of G at each point. For each s_{ij}, we can cover U_i with a finite number of open sets

$U_{ij\ell}$ such that each $s_{ij}|_{U_{ij\ell}}$ comes from a section of a suitable $G_{k(ij\ell)}$ over $U_{ij\ell}$. Doing this for all the s_{ij}, and taking $k = \max(k(ij\ell))$, we find that all the s_{ij} come from sections of G_k. Hence $G_k = G$, and so our sequence is stationary.

Corollary 7.9. Let X be a locally noetherian prescheme. Then any direct sum of injective $\mathcal{O}_X$-modules is injective.

Corollary 7.10. Let X be a locally noetherian prescheme. Then any injective $\mathcal{O}_X$-module can be written uniquely as a direct sum of indecomposable injective $\mathcal{O}_X$-modules.

Definition. Let X be a locally noetherian prescheme, let $x \in X$, and let x' be a specialization of x, i.e., $x' \in \{x\}^-$. Let I be an injective hull of $k(x)$ over the local ring $\mathcal{O}_{x,X}$. We define $J(x,x')$ to be the restriction of the sheaf $i_x(I)$ to the closed subset $\{x'\}^-$ of $\{x\}^-$, as in [G, II.Thm. 2.9.1]. If $x = x'$, we write simply $J(x)$ for $J(x,x) = i_x(I)$. Note that $J(x,x')$ is an indecomposable injective $\mathcal{O}_X$-module.

Theorem 7.11. Let X be a locally noetherian prescheme. Then the indecomposable injective $\mathcal{O}_X$-modules are precisely the sheaves $J(x,x')$ defined above, for every pair of points x specializing to x' of X.

<u>Lemma 7.12</u>. Let X be a locally noetherian prescheme, and let I be an injective $\mathcal{O}_X$-module. Then the stalk I_x of I at each point $x \in X$ is an injective module over the local ring $\mathcal{O}_x$ of x.

<u>Proof</u>. Since $\mathcal{O}_x$ is a noetherian ring, it is sufficient to show that whenever $M \subseteq N$ is an inclusion of $\mathcal{O}_x$-modules of finite type, and $\varphi\colon M \longrightarrow I_x$ is a map, then φ extends to N. We can find coherent sheaves $\overline{M} \subseteq \overline{N}$ on X, with stalks M and N at x [EGA.I.9.4.8], and we can find a map of $\overline{M}$ to I in a suitable neighborhood U of x, extending φ, since $\overline{M}$ is a sheaf of finite presentation [EGA 0.5.2.6]. This gives a map of $\overline{M}_U$ to I. Now since I is injective, this extends to a map of $\overline{N}_U$ to I, whose stalk at x extends φ.

<u>Proof of theorem</u>. Let I be an indecomposable injective $\mathcal{O}_X$-module. We define <u>Supp(I)</u> to be the set of points $x \in X$ such that the stalk I_x is non-zero. Let $x' \in \mathrm{Supp}(I)$ be a maximal point (i.e., one which is not a specialization of any other point in Supp(I).) Then by the lemma, the stalk $I_{x'}$ is an injective $\mathcal{O}_{x'}$-module. Let I_o be an indecomposable injective direct summand of $I_{x'}$. Then by Proposition 7.4, I_o is of the form $I(\mathfrak{p})$ for some

prime ideal $\mathfrak{p}$ of $\mathcal{O}_{x'}$. Let $\mathfrak{p}$ correspond to the point $x \in X$. Then $i_x(I(\mathfrak{p})) = J(x)$, using the notation above. Furthermore, there is a natural inclusion of $\mathcal{O}_Z$ in $J(x)$, where $Z = \{x\}^-$. Since the stalk $J(x)_{x'} = I_o$ is mapped into $I_{x'}$, we have a map $\mathcal{O}_{Z,x'} \longrightarrow I_{x'}$, and hence can find a map $\varphi\colon \mathcal{O}_Z \rightarrow I$ extending it in a suitable neighborhood U of x'. Let $Z' = \{x'\}^-$. Then the image of φ has support in Z', since x' was chosen maximal in Supp(I), and by construction, φ is injective at x'. Now applying Lemma 7.7 to the kernel of φ, we see that φ factors through $\mathcal{O}_{Z,Z'}$, the restriction of $\mathcal{O}_Z$ to the closed subset Z', to give a map $\psi\colon \mathcal{O}_{Z,Z'} \rightarrow I$ (defined on U), and that by shrinking U a bit, we may assume that ψ is <u>injective</u> on U. In other words, we have an injection of sheaves on X,

$$\psi_U\colon \mathcal{O}_{Z,Z',U} \longrightarrow I$$

where $\mathcal{O}_{Z,Z',U}$ is the restriction of $\mathcal{O}_{Z,Z'}$ to the open set U.

Now it is easily seen that $J(x,x')$ is an injective hull of $\mathcal{O}_{Z,Z',U}$, so ψ_U extends to give a map $J(x,x') \longrightarrow I$, which is necessarily an inclusion since ψ_U is. Now $J(x,x')$ being injective, is a direct summand, so must be equal to I since I was indecomposable.

q.e.d.

<u>Corollary 7.13</u>. Let X be a locally noetherian prescheme. Then every injective $\mathcal{O}_X$-module is uniquely a direct sum of injectives $J(x,x')$ defined above.

<u>Corollary 7.14</u>. Let A be a noetherian ring, let I be an injective A-module, and let $X = \operatorname{Spec} A$. Then $\tilde{I}$ is an injective $\mathcal{O}_X$-module.

<u>Proof</u>. Follows from Theorem 7.3, Propositions 7.4 and 7.6, and Corollary 7.9.

<u>Corollary 7.15</u>. Let A be a noetherian ring, let M be an A-module of finite type, and let N be any A-module. Then

$$\underline{\operatorname{Ext}}^i_X(\tilde{M},\tilde{N}) = \operatorname{Ext}^i_A(M,N)^\sim$$

for all i, where $X = \operatorname{Spec} A$.

<u>Proof</u>. Since M is of finite type,

$$\underline{\operatorname{Hom}}(\tilde{M},\tilde{N}) = \operatorname{Hom}(M,N)^\sim$$

for any N. Take an injective resolution $I^\cdot$ of N. Then $\tilde{I}^\cdot$ is an injective resolution of $\tilde{N}$ by the previous Corollary. Since $\sim$ is an exact functor, we have the result.

Lemma 7.16. Let X be a prescheme, and let F be an $\mathcal{O}_X$-module. Then F is injective in $\mathrm{Mod}(X)$ if and only if there is an open cover $\{U_\alpha\}$ of X such that for each α, $F|_{U_\alpha}$ is injective in the category $\mathrm{Mod}(U_\alpha)$.

Proof. If F is injective, then F restricted to any open set is injective. Indeed, given $G' \subseteq G$ on U and a map $G' \longrightarrow F|_U$, we deduce a map $G'^X \longrightarrow F$, where G'^X is the sheaf G' extended by zero outside U. This extends to a map $G^X \longrightarrow F$, hence $G \longrightarrow F|_U$.

On the other hand, to test whether F is injective, it is enough, by Zorn's lemma, to show that for every sheaf G in a family of generators of the category, and for every map $\varphi: G' \longrightarrow F$ of a subsheaf G' of G to F, φ extends to G. Since the sheaves $\mathcal{O}_U$ (which is $\mathcal{O}_X$ on U and 0 outside), for U arbitrarily small, form a family of generators of $\mathrm{Mod}(X)$, we see that the question of injectivity is local, as required.

Proposition 7.17. Let X be a locally noetherian prescheme, and let F be a quasi-coherent $\mathcal{O}_X$-module. Then the following conditions are equivalent:

(i) F is an injective $\mathcal{O}_X$-module.

(ii) F is isomorphic to a direct sum of sheaves $J(x)$ for various $x \in X$.

(iii) For every $x \in X$, the stalk F_x of F at x is an injective $\mathcal{O}_x$-module.

(iv) For all coherent sheaves G on X, $\underline{\mathrm{Ext}}^1(G,F) = 0$.

(v) There is an open cover $\{U_\alpha\}$ of X such that $F|_{U_\alpha}$ is an injective $\mathcal{O}_{U_\alpha}$-module for all α.

Proof. (i) $\Longrightarrow$ (ii) We know by Corollary 7.13 that F is isomorphic to a direct sum of $J(x,x')$. Since F is quasi-coherent, each $J(x,x')$ must be also. But $J(x,x')$ is quasi-coherent if and only if $x = x'$, i.e., $J(x,x') = J(x)$.

(ii) $\Longrightarrow$ (iii) Clear.

(iii) $\Longrightarrow$ (iv) Since G is coherent, $\underline{\mathrm{Ext}}$ commutes with passage to stalks, and all the stalks are zero.

(iv) $\Longrightarrow$ (v) On a noetherian affine $U = \operatorname{Spec} A$ of X $\underline{\mathrm{Ext}}^1(G,F)|_U = \mathrm{Ext}_A^1(M,N)^\sim$ where $G = \tilde{M}$ and $F = \tilde{N}$. The result then follows from [EGA I.9.4.8], Corollary 7.14, and the well-known fact that on a noetherian ring, N is injective if and only if $\mathrm{Ext}_A^1(M,N) = 0$ for all A-modules M of finite type.

(v) $\Longrightarrow$ (i) by the lemma.

<u>Theorem 7.18</u>. Let X be a locally noetherian prescheme. Then every quasi-coherent $\mathcal{O}_X$-module F can be embedded in a quasi-coherent, injective $\mathcal{O}_X$-module I.

<u>Proof</u>. Indeed, we will show that the injective hull I of a quasi-coherent sheaf F is quasi-coherent. Let I be written as a direct sum of sheaves $J(x,x')$, by Corollary 7.13. For each such $J(x,x')$, since I is an injective hull of F, there is a neighborhood U of x', and a section $s \in J(x,x')(U)$ which is also a section of $F(U)$. By shrinking U if necessary, we may assume that the support of s is just $U \cap Z'$, where $Z' = \{x'\}^-$. Since s is a section of a quasi-coherent sheaf F, it must be annihilated by some power of the ideal $\mathcal{O}_{Z'}$ of Z'. This implies $x = x'$, for if $x \neq x'$, no section of $J(x,x')$ is annihilated by any power of $\mathcal{O}_{Z'}$. Thus $J(x,x') = J(x)$ is quasi-coherent, and so I is quasi-coherent.

<u>Corollary 7.19</u>. Let X be a locally noetherian prescheme. Then the natural functor

$$D^+(Qco(X)) \longrightarrow D^+_{qc}(X)$$

is an equivalence of categories.

<u>Proof</u>. (Cf. §1 for notations). Follows from [I.4.8].

As an application, we give the following result on complexes of finite injective dimension.

Proposition 7.20. Let X be a locally noetherian prescheme, let $A = \mathrm{Mod}(X)$, and let $F^{\cdot} \in \mathrm{Ob}\, K^{+}_{qc}(A)$. Then the equivalent conditions (i),(ii), and (iii) of [I.7.6] are also equivalent to the following:

$(i)_{qc}$ $F^{\cdot}$ admits a quasi-isomorphism $F^{\cdot} \longrightarrow I^{\cdot}$ into a bounded complex of quasi-coherent injective $\mathcal{O}_X$-modules.

(ii) The functor $\underline{\underline{R}}\, \underline{\mathrm{Hom}}^{\cdot}(\cdot, F^{\cdot})$ from $D(A)^{*}$ to $D(A)$ is way-out left.

(iii) There is an integer n_o such that $\underline{\mathrm{Ext}}^{i}(G, F^{\cdot}) = 0$ for all $G \in \mathrm{Mod}(X)$ and all $i > n_o$.

$(iii)_c$ There is an integer n_o such that $\underline{\mathrm{Ext}}^{i}(G, F^{\cdot}) = 0$ for all $G \in \mathrm{Coh}(X)$ and all $i > n_o$.

Proof. $(i)_{qc} \Longrightarrow (i) \Longrightarrow (ii) \Longrightarrow (iii) \Longrightarrow (iii)_c$ are all easy as before. It remains only to prove $(iii)_c \Longrightarrow (i)_{qc}$. This is similar to the proof of (iii) $\Longrightarrow$ (i) in loc. cit. using Proposition 7.17(iv) and Theorem 7.18 above.

Example. If X is a locally noetherian prescheme, the category $Qco(X)$ of quasi-coherent sheaves on X may not be locally noetherian. Thus we do not know the structure of injectives in that category, and we do not know whether every injective object of $Qco(X)$ is injective in $Mod(X)$.

Here is the example. Let X_1 be the projective plane, let E_1 be a line in X_1, and let x_1 be a closed point of E_1. Having defined X_n, E_n, x_n, define X_{n+1} to be X_n blown up at the point x_n, let E_{n+1} be the exceptional curve, and let x_{n+1} be a closed point of E_{n+1}. Define

$$X = \coprod_{n=1}^{\infty} (X_n - x_n) ,$$

where we glue $X_n - x_n$ to the open subset $X_{n+1} - E_{n+1}$ of $X_{n+1} - x_{n+1}$. Then X is an integral, locally noetherian scheme.

However, $Qco(X)$ is not a locally noetherian category. Indeed, let $\mathcal{I}$ be a non-zero sheaf of ideals of $\mathcal{O}_X$, and let F be a noetherian generator of the category which admits a map into $\mathcal{O}_X$ not factoring through $\mathcal{I}$. Then the image of F must be a noetherian non-zero ideal G of $\mathcal{O}_X$.

For each $k = 1,2,\ldots$, let Y_k be a closed subset of X,

$$Y_k = \bigsqcup_{n=k}^{\infty} (E_n - x_n) .$$

Then $Y_1 > Y_2 > \cdots$. Let $\mathcal{J}_k$ be the sheaf of ideals of Y_k. Then $G\mathcal{J}_1 < G\mathcal{J}_2 < \cdots$ which is a contradiction to the statement G is noetherian.

We conclude that the category $\mathrm{Qco}(X)$ does not have a family of noetherian generators.

<u>Remark</u>. We do not know if the analogue of Corollary 7.19 is true for unbounded complexes, i.e., whether the natural functor

$$D(\mathrm{Qco}(X)) \longrightarrow D_{qc}(X)$$

is an equivalence of categories. However, we conjecture it to be true when X is a <u>regular</u> noetherian scheme of finite Krull dimension, because in that case the category $\mathrm{Mod}(X)$ has finite injective dimension.

CHAPTER III. DUALITY FOR PROJECTIVE MORPHISMS

§1. Differentials.

In this section we recall some facts on relative differentials which we will need in the sequel. These results will surely be in [EGA] eventually, but for the moment the best reference seems to be [SGA 60-61, exposé II].

Definition. If $A \longrightarrow B$ is a morphism of rings, and M a B-module, we define $\mathrm{Der}_A(B,M)$ to be the A-module of derivations of B into M over A. We define $\Omega^1_{B/A}$, the module of relative one-differentials of B over A, to be the B-module representing the functor

$$M \longrightarrow \mathrm{Der}_A(B,M).$$

In other words, there is a derivation $d\colon B \longrightarrow \Omega^1_{B/A}$ given, such that for any B-module M, the natural map

$$\mathrm{Hom}_B(\Omega^1_{B/A},M) \longrightarrow \mathrm{Der}_A(B,M)$$

is an isomorphism.

If $f\colon X \longrightarrow Y$ is a morphism of preschemes, we define $\Omega^1_{X/Y}$, the sheaf of relative one-differentials of X over Y, by considering open affines in X and Y, and glueing the corresponding modules $\Omega^1_{B/A}$.

Definition. [EGA IV 6.8.1] A morphism $f: X \to Y$ of preschemes is smooth if it is flat, locally of finite presentation, and for every $y \in Y$, the fibre $f^{-1}(y)$ is locally noetherian, and geometrically regular (i.e., "absolutely non-singular").

Examples. 1. An open immersion is smooth.

2. A composition of smooth morphisms is smooth.

3. Smooth morphisms are stable under base extensions.

4. A prescheme X over a field k is smooth $\Longleftrightarrow$ it is locally noetherian and geometrically regular.

Proposition 1.1. [SGA 60-61, II.4.3] Let $f: X \to Y$ be a smooth morphism of preschemes over another prescheme S. Then $\Omega^1_{X/Y}$ is locally free (of rank n = relative dimension of X over Y), and there is an exact sequence

$$0 \to f^*(\Omega^1_{Y/S}) \to \Omega^1_{X/S} \to \Omega^1_{X/Y} \to 0.$$

Definition. A closed subscheme Y of a prescheme X is locally a complete intersection if every point $y \in Y$ has a neighborhood U such that in U, the ideal J_Y of Y is generated by an $\mathcal{O}_X$-sequence, i.e., a collection of sections $s_1,\dots,s_r$ such that s_1 is a non-zero divisor in $\mathcal{O}_X$, and for each

$i = 2,\ldots,r$, s_i is a non-zero divisor in $\mathcal{O}_X/(s_1,\ldots,s_{i-1})$.

Proposition 1.2. [SGA 60-61 II.4.10] Let X be a locally noetherian prescheme, smooth over a locally noetherian prescheme S, and let Y be a closed subprescheme of X. Then the following conditions are equivalent:

(i) Y is smooth over S

(ii) $\Omega^1_{Y/S}$ is locally free, and the sequence

$$0 \longrightarrow J/J^2 \longrightarrow i^*\Omega^1_{X/S} \longrightarrow \Omega^1_{Y/S} \longrightarrow 0$$

is exact, where J is the sheaf of ideals of Y, and $i: Y \longrightarrow X$ the immersion.

Furthermore, in that case Y is locally a complete intersection in X.

Definition. Let X be a prescheme, and let

$$(*) \qquad 0 \longrightarrow E' \longrightarrow E \longrightarrow E'' \longrightarrow 0$$

be an exact sequence of locally free sheaves of ranks r, r+s, and s, respectively. Then we define an isomorphism

$$\varphi(*): \Lambda^{r+s}E \longrightarrow \Lambda^r E' \otimes \Lambda^s E''$$

as follows. Choose a basis $e_1,\ldots,e_{r+s}$ of E locally such that $e_1,\ldots,e_r$ form a basis of E', and the images $\bar{e}_j$ of $e_{r+1},\ldots,e_{r+s}$

form a basis of E". Then map

$$e_1\wedge\ldots\wedge e_{r+s} \longrightarrow (e_1\wedge\ldots\wedge e_r)\otimes(\bar{e}_{r+1}\wedge\ldots\wedge\bar{e}_{r+s}).$$

Remark. We have made a choice here, and it is emphatically not true that all imaginable compatibilities of these isomorphisms $\varphi(*)$ hold. For example, if $(\check{*})$ is the dual exact sequence to $(*)$, then $\varphi(\check{*})$ and $\varphi(*)^{\vee}$ commute only to within a factor of $(-1)^{rs}$.

Lemma 1.3. Let $0 \subseteq E_1 \subseteq E_2 \subseteq E$ be locally free sheaves on a prescheme X. Then the four exact sequences

$$0 \longrightarrow E_1 \longrightarrow E_2 \longrightarrow E_2/E_1 \longrightarrow 0$$

$$0 \longrightarrow E_1 \longrightarrow E \longrightarrow E/E_1 \longrightarrow 0$$

$$0 \longrightarrow E_2 \longrightarrow E \longrightarrow E/E_2 \longrightarrow 0$$

$$0 \longrightarrow E_2/E_1 \longrightarrow E/E_1 \longrightarrow E/E_2 \longrightarrow 0$$

give rise to a commutative diagram of isomorphisms φ among their highest exterior powers.

Proof. Left to reader.

Definition. a) Let $f\colon X \longrightarrow Y$ be a smooth morphism of relative dimension n. Then we define $\omega_{X/Y} = \Lambda^n\Omega^1_{X/Y}$. Note by Proposition 1.1 that $\omega_{X/Y}$ is a locally free sheaf of rank one.

b) Let $f: X \longrightarrow Y$ be a closed immersion which is locally a complete intersection of codimension n (i.e., locally defined by n equations), and let J be the sheaf of ideals of X. Then we define $\omega_{X/Y} = (\Lambda^n(J/J^2))^{\vee}$, where $^{\vee}$ denotes dual. Note that J/J^2 is locally free of rank n on X, so that $\omega_{X/Y}$ is a locally free sheaf of rank one on Y.

Remarks. 1. Note if $f: X \rightarrow Y$ is smooth and a closed immersion, then it is locally an isomorphism, so the two definitions coincide.

2. If $f: X \rightarrow Y$ is either smooth, or a locally complete intersection, and if $Y' \rightarrow Y$ is a base change, and $X' = X \times_Y Y'$, then $pr_1^*(\omega_{X/Y}) = \omega_{X'/Y'}$. This follows from the fact that differentials and ideals of subschemes are compatible with base extension.

Lemma 1.4. Let $X \xrightarrow{f} Y \xrightarrow{g} Z$ be morphisms of preschemes, with g smooth. Let $\Gamma: X \longrightarrow X \times_Z Y$ be the graph morphism. Then Γ is locally a complete intersection, and $\omega_{X/X\times_Z Y} = f^*\omega_{Y/Z}^{\vee}$.

Proof. We apply Proposition 1.2 to $W = X \times_Z Y$, X, and the section Γ. Note that $p_1: W \rightarrow X$ is smooth by base extension from g, hence Γ is a local complete intersection, and

$J/J^2 \cong \Gamma^*\Omega^1_{W/X}$. Therefore $\omega_{X/W} = \Gamma^*\check{\omega}_{W/X}$. But again by base extension, $\omega_{W/X} = p_2^*\omega_{Y/Z}$, and $f = p_2\Gamma$, so $\omega_{X/W} = f^*\check{\omega}_{Y/Z}$ as required.

<u>Definition 1.5</u>. Let $X \xrightarrow{f} Y \xrightarrow{g} Z$ be morphisms of locally noetherian preschemes, and suppose that f,g, and gf is each either smooth or a local complete intersection. Then we define an isomorphism

$$\zeta_{f,g}\colon \omega_{X/Z} \xrightarrow{\sim} f^*\omega_{Y/Z} \otimes \omega_{X/Y}.$$

There are four cases to consider.

a) f,g, and gf are all smooth. Then we take ζ to be φ of the exact sequence of Proposition 1.1.

b) f,g, and gf are all local complete intersections. If J is the ideal of Y in Z, and K is the ideal of X in X, then we have an exact sequence on X,

$$0 \longrightarrow f^*(J/J^2) \longrightarrow K/K^2 \longrightarrow (K/J)/(K/J)^2 \longrightarrow 0.$$

We take φ of this exact sequence, then dualize, and take the inverse isomorphism to be ζ.

c) f is a local complete intersection, with g and gf smooth. We take φ of the exact sequence of Proposition 1.2, tensor with $\omega_{X/Y}$, and take the inverse to get ζ.

d) f and gf local complete intersections, and g smooth. Then by the lemma above, Γ is a local complete intersection, and we can use b) above applied to Γ and p_2 to obtain

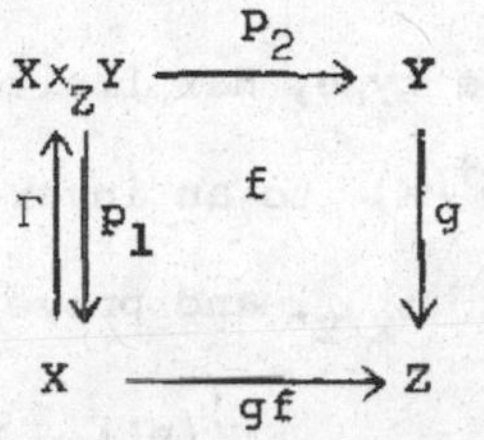

$$\zeta_{\Gamma,p_2}: \omega_{X/Y} \xrightarrow{\sim} \Gamma^*\omega_{X\times_Z Y/Y} \otimes \omega_{X/X\times_Z Y}.$$

Now by base extension, $\omega_{X\times_Z Y/Y} = p_1^*\omega_{X/Z}$, and by the lemma, $\omega_{X/X\times_Z Y} = f^*\omega_{Y/Z}^{\vee}$. Transposing and taking the inverse, we obtain

$$\zeta_{f,g}: \omega_{X/Z} \xrightarrow{\sim} f^*\omega_{Y/Z} \otimes \omega_{X/Y}.$$

Proposition 1.6. Let $X \xrightarrow{f} Y \xrightarrow{g} Z \xrightarrow{h} W$ be three morphisms of locally noetherian preschemes, and suppose that each of the morphisms f,g,h,gf,hg,hgf is either smooth or a local complete intersection. Then the isomorphisms ζ give a commutative diagram

$$\zeta_{h,g}\zeta_{f,hg} = \zeta_{f,g}\zeta_{gf,h}\ .$$

Proof. Use Lemma 1.3.

*Remark. The reader will realize later that the proper context for the notion of $\omega_{X/Y}$ just studied is that of a Gorenstein morphism, and we will leave him to elaborate on the following indications. A morphism $f\colon X \longrightarrow Y$ of locally noetherian preschemes is called Gorenstein if it is locally of finite type, has finite Tor-dimension, and if $f^!(\mathcal{O}_Y)$ is isomorphic in $D^+(X)$ to an invertible sheaf. Then we call that invertible sheaf $\omega_{X/Y}$, and prove that

$$f^!(F^\cdot) = \underline{\underline{L}}f^*(F^\cdot) \otimes \omega_{X/Y}$$

for all $F^\cdot \in D^+_{qc}(Y)$.

Smooth morphisms, and locally complete intersection morphisms are Gorenstein, and in those two cases the sheaf $\omega_{X/Y}$ is the one we have already defined. Furthermore, if $f\colon X \longrightarrow Y$ and $g\colon Y \longrightarrow Z$ are Gorenstein, so is gf, and there is an isomorphism

$$\zeta_{f,g}\colon \omega_{X/Z} \xrightarrow{\sim} f^*\omega_{Y/Z} \otimes \omega_{X/Z}.$$

For a composition of three Gorenstein morphisms, there is a commutative diagram as in the Proposition.$_*$

§2. $f^{\#}$ for a smooth morphism f.

Definition. Let $f\colon X \longrightarrow Y$ be a smooth morphism of preschemes. Then we define a functor

$$f^{\#}\colon\ D(Y) \longrightarrow D(X)$$

by $$f^{\#}(G^{\cdot}) = f^{*}(G^{\cdot}) \otimes \omega_{X/Y}[n],$$

where $[n]$ means "shift n places to the left". Observe that f is flat, so $f^{*} = \underline{\underline{L}}f^{*}$ is defined on all of $D(Y)$, and $\omega_{X/Y}$ is an invertible sheaf on X, hence is an element of $D(X)_{fTd}$, so that the tensor product is defined [II §4].

Proposition 2.1. Let $f\colon X \longrightarrow Y$ be a smooth morphism, let $u\colon Y' \longrightarrow Y$ be a morphism of finite Tor-dimension, and let $X' = X \times_Y Y'$. Then there is a natural isomorphism

$$\begin{array}{ccc} X' & \xrightarrow{v} & X \\ {\scriptstyle g}\downarrow & & \downarrow{\scriptstyle f} \\ Y' & \xrightarrow{u} & Y \end{array}$$

$$(\underline{\underline{L}}v^{*})f^{\#} = g^{\#}\,\underline{\underline{L}}u^{*}$$

of functors from $D(Y)$ to $D(X')$.

Proof. This follows from [II 5.4] and [II 5.9] modified to include the case of finite Tor-dimension, and the compatibility of $\omega_{X/Y}$ with base extension [§1 above].

Remark. Following the conventions of [II §6], we write "=" instead of naming the isomorphism and keeping track of it. However, in the following Proposition we do not write "=", because the isomorphism depends on a choice of sign made in §1 above. In general, we will write "=" below when there can be no doubt about the isomorphism being compatible with all previous ones, and we will name those isomorphisms where there may be a question of sign, or of choice of coordinates, etc.

Proposition 2.2. Let $f: X \longrightarrow Y$ and $g: Y \longrightarrow Z$ be two smooth morphisms. Then there is an isomorphism

$$\zeta_{f,g}: (gf)^{\#} \xrightarrow{\sim} f^{\#}g^{\#}$$

of functors from $D(Z)$ to $D(X)$. Furthermore, for a composition of three smooth morphisms the isomorphisms ζ give a commutative diagram.

Proof. We define $\zeta_{f,g}$ using the ζ of Definition 1.5 above, and the isomorphisms [II 5.4] and [II 5.9]. The compatibility then follows from Proposition 1.6.

Proposition 2.3. Let $f: X \longrightarrow Y$ be a smooth morphism. Then $f^{\#}$ takes $D_{qc}(Y)$ to $D_{qc}(X)$, and, if X and Y are locally noetherian, it takes $D_c(Y)$ to $D_c(X)$.

Proof. Obvious.

<u>Proposition 2.4</u>. Let $f: X \longrightarrow Y$ be a smooth morphism. Then

a) There is a functorial isomorphism

$$f^{\#}(F^{\cdot} \underline{\underline{\otimes}} G^{\cdot}) \xrightarrow{\sim} f^{\#}(F^{\cdot}) \underline{\underline{\otimes}} f^{*}(G^{\cdot})$$

provided either $F^{\cdot}, G^{\cdot} \in D^{-}(Y)$, or one of $F^{\cdot}, G^{\cdot}$ is in $D^{b}(Y)_{fTd}$, and the other is in $D(Y)$.

b) There is a functorial homomorphism

$$f^{\#}(\underline{\underline{R}}\ \underline{Hom}^{\cdot}(F^{\cdot}, G^{\cdot})) \longrightarrow \underline{\underline{R}}\ \underline{Hom}^{\cdot}(f^{*}F^{\cdot}, f^{\#}G^{\cdot})$$

for $F^{\cdot} \in D(Y)$ and $G^{\cdot} \in D^{+}(Y)$. It is an isomorphism if Y is locally noetherian, and $F^{\cdot} \in D_{c}^{-}(Y)$.

<u>Proof</u>. Left to reader. (Use [II 5.8],[II 5.9],[II 5.13], and [II 5.16].)

§3. Recall of the Explicit Calculations.

In this section we recall the calculations of the cohomology of projective space, as done in [EGA III §1]. First we must define the Cech resolution of a sheaf, and we follow [G, II §5].

Let X be a prescheme, let $\mathcal{U} = (U_i)$ be a family of open sets of X, and let F be an $\mathcal{O}_X$-module. Then we define the Cech complex of F, $C^{\cdot}(\mathcal{U},F)$, as follows.

For each $p \geq 0$, and for each $(p+1)$-tuple of indices $i_o < \cdots < i_p$ let $U_{i_o,\ldots,i_p} = U_{i_o} \cap \cdots \cap U_{i_p}$. Define the sheaf $C^p(\mathcal{U},F)$ by giving its sections on an open set V as follows:

$$C^p(\mathcal{U},F)(V) = \prod_{i_o < \cdots < i_p} F(V \cap U_{i_o,\ldots,i_p}).$$

One checks easily that this is a sheaf. In fact, it is the product, over all $i_o < \cdots < i_p$, of the sheaves $i_*(F|_{U_{i_o,\ldots,i_p}})$, where $i\colon U_{i_o,\ldots,i_p} \longrightarrow X$ is the inclusion. If

$$\alpha \in C^p(\mathcal{U},F)(V)$$

is a section, we represent it by its components,

$$\alpha = \prod \alpha_{i_o,\ldots,i_p}$$

with

$$\alpha_{i_o,\ldots,i_p} \in F(V \cap U_{i_o,\ldots,i_p}).$$

We define the boundary map

$$d\colon C^p(\mathcal{U},F) \longrightarrow C^{p+1}(\mathcal{U},F)$$

as follows. If $\alpha \in C^p(\mathcal{U},F)(V)$ is a section, as above, then the components of $d\alpha$ are given by

$$(d\alpha)_{i_o,\ldots,i_{p+1}} = \sum (-1)^j \rho_j \alpha_{i_o,\ldots,\hat{i}_j,\ldots,i_{p+1}}$$

where ρ_j is the appropriate restriction map on sections of F.

Finally, we define an augmentation

$$\epsilon\colon F \longrightarrow C^o(\mathcal{U},F)$$

by sending a section $\alpha \in E(V)$ to the product of its restrictions $\alpha_i \in F(V \cap U_i)$.

Proposition 3.1. [G, II.5.2.1] Suppose that $\mathcal{U}$ is a covering of X. Then the augmentation ϵ gives a quasi-isomorphism of F to the Cech complex $C^{\cdot}(\mathcal{U},F)$ of F (i.e., it is a "resolution" of F, in the old language).

Proposition 3.2. Let $f\colon X \longrightarrow Y$ be a separated morphism of preschemes, let $\mathcal{U} = (U_i)$ be a family of open subsets of X such that $f|_{U_i}$ is an affine morphism for each i, and let F be a quasi-coherent $\mathcal{O}_X$-module. Then the sheaves $C^p(\mathcal{U},F)$ are f_*-acyclic.

Proof. Since a product of f_*-acyclic sheaves is f_*-acyclic, we need only show that if U is an open subset of X such that $f|_U$ is an affine morphism, and if $i\colon U \longrightarrow X$ is the inclusion, then $i_*(F)$ is f_*-acyclic. Now since f is separated, and $f|_U$ is affine, it follows that i is an affine morphism. On the other hand, since F is quasi-coherent, $F|_U$ is acyclic for the affine morphisms i and fi, by [EGA III 1.3.2]. Hence by the spectral sequence of derived functors [II.5.1], $i_*(F|_U)$ is f_*-acyclic.

Corollary 3.3. Let $f\colon X \longrightarrow Y$ be a separated morphism of preschemes, let $\mathcal{U} = (U_i)$ be an open cover of X such that $f|_{U_i}$ is an affine morphism for each i, and let F be a quasi-coherent sheaf on X. Then the natural maps

$$f_*(C^{\cdot}(\mathcal{U},F)) \xrightarrow{\ \xi\ } \underline{\underline{R}}f_*(C^{\cdot}(\mathcal{U},F)) \xleftarrow{\ \underline{\underline{R}}f_*(\epsilon)\ } \underline{\underline{R}}f_*(F)$$

are isomorphisms in $D(Y)$. (Here ξ is the canonical map in the definition of the derived functor, cf. [I.5].)

Proof. Follows from the two previous results and from [I.5.1] and [I.5.3β].

Now we will apply these results to projective space. Let Y be a prescheme, and let $X = \mathbb{P}^n_Y$ be the n-dimensional projective space over Y, i.e., $X = \underline{\text{Proj}}\ \mathcal{O}_Y[T_o,\cdots,T_n]$ where the T_i are indeterminates. For each i, let $U_i = X_{T_i}$, the place where $T_i \neq 0$. Then $\mathcal{U} = (U_i)$ is a finite open cover of X, and $f|_{U_i}$ is an affine morphism for each i, where $f: X \longrightarrow Y$ is the projection. Indeed, $U_i \cong \mathbb{A}^n_Y$, affine n-space.

On U_o we fix a set of inhomogeneous coordinates

$$t_i = T_i/T_o, \qquad i = 1,\cdots,n.$$

Let $\omega = \omega_{X/Y}$ be the relative n-differential forms on X over Y. (It is well known that one can find an isomorphism $\omega \cong \mathcal{O}_X(-n-1)$, but we will not use this isomorphism, because of its non-intrinsic nature.) Then

$$\tau = dt_1 \wedge \cdots \wedge dt_n$$

is a generating section of $\omega|_{U_o}$, since $\Omega^1_{U_o/Y}$ is free of rank n, and generated by $dt_1,\cdots,dt_n$. Since $\omega(n+1) \cong \mathcal{O}_X$, τ extends to a global section

$$\tau \in \Gamma(X,\omega(n+1))$$

which we will also call τ.

Multiplication by $T_o \cdots T_n$ gives a map from ω to $\omega(n+1)$, which is an isomorphism on $U_{o,\ldots,n}$, so we can consider the section

$$\tau/T_o \cdots T_n \in \omega(U_{o,\ldots,n}) .$$

This section is an n-cocycle in the complex $f_*(C^\cdot(\mathfrak{U},\omega))$, and so using Corollary 3.3 above, defines an element

$$\bar{\tau} \in \Gamma(Y,R^n f_*(\omega)).$$

<u>Theorem 3.4</u>. [EGA III 2.1.12] Let Y be a prescheme, let $X = \mathbb{P}^n_Y$, let f be the projection, and let $\omega = \omega_{X/Y}$ be the relative n-differentials. Then $R^n f_*(\omega)$ is an invertible sheaf on Y, and $\bar{\tau}$ is a generating section, hence it defines an isomorphism

$$\gamma: \; R^n f_*(\omega) \xrightarrow{\sim} \mathcal{O}_Y$$

by sending $\bar{\tau}$ to 1. Furthermore,

$$R^i f_*(\mathcal{O}_X(m)) = 0 = R^{n-i} f_*(\omega(-m))$$

for $0 < i < n$, $m \in \mathbb{Z}$, and for $i = 0$, $m < 0$, and the cup-product

$$f_*(\mathcal{O}_X(m)) \times R^n f_*(\omega(-m)) \longrightarrow R^n f_*(\omega)$$

is a perfect pairing of locally free sheaves for all $m \geq 0$.

Remarks. 1. Note that the isomorphism γ we have constructed above is compatible with arbitrary base extension, since everything in the construction is flat over Y.

2. It is natural to ask whether the isomorphism γ is stable under automorphisms of the projective space, and we will see later (Corollary 10.2) that indeed it is.

§4. The trace map for projective space.

In this section we define the trace isomorphism

$$\mathrm{Trp}_f:\ \underline{\underline{R}}f_*f^{\#}(G^\cdot) \xrightarrow{\ \sim\ } G^\cdot$$

for $G^\cdot \in D^+_{qc}(Y)$, where Y is a locally noetherian prescheme, and $f: \mathbb{P}^n_Y \longrightarrow Y$ is the projection. The definition uses the results of [II.7] on locally noetherian preschemes. Without using these results, we can define the trace map only for $G^\cdot \in D^b_{qc}(Y)$.

Lemma 4.1. Let Y be a locally noetherian prescheme, let $X = \mathbb{P}^n_Y$, and let $f: X \longrightarrow Y$ be the projection. Then every quasi-coherent sheaf F on X is a quotient of a sheaf of the form

$$L = \oplus\, f^*(G_i)(-m_i)$$

where the G_i are quasi-coherent sheaves on Y, and the $m_i > 0$ are integers.

Proof. Since F is the direct limit of its coherent subsheaves, we may assume that F itself is coherent. For each $m > 0$ we have a natural map

$$f^*f_*(F(m)) \longrightarrow F(m),$$

and we know from Serre's theorem [EGA III 2.2.1] that for each noetherian open subset $V \subseteq Y$, the restriction of this map to

$f^{-1}(V)$ is surjective for large enough m. Hence the map

$$\bigoplus_{m>0} f^*f_*(F(m))(-m) \longrightarrow F$$

is surjective on all of X, so we are done.

Lemma 4.2. For any sheaf L on X of the form of the previous lemma, $R^if_*(L) = 0$ for $i \neq n$.

Proof. It is sufficient to show that for G quasi-coherent on Y, and $m > 0$, $R^if_*(f^*(G)(-m)) = 0$ for $i \neq n$. The question is local on Y, so we may assume Y quasi-compact, and use the projection formula [II.5.6] (note $f^*(G)(-m) = f^*(G) \otimes \mathcal{O}_X(-m)$):

$$\underline{\underline{R}}f_*(f^*(G)(-m)) \xrightarrow{\sim} \underline{\underline{R}}f_*(\mathcal{O}_X(-m)) \underline{\underline{\otimes}} G.$$

But $R^if_*(\mathcal{O}_X(-m)) = 0$ for $i \neq n$ by the explicit calculations (Theorem 3.4), and $R^nf_*(\mathcal{O}_X(-m))$ is locally free on Y, so there is only one non-zero sheaf on the right, and we are done.

Proposition 4.3. Let Y be a locally noetherian prescheme, let $X = \mathbb{P}^n_Y$, and let $f: X \to Y$ be the projection. Then there is a functorial isomorphism

$$Trp_f: \underline{\underline{R}}f_*f^{\#}(G^{\cdot}) \xrightarrow{\sim} G^{\cdot}$$

for $G^{\cdot} \in D^+_{qc}(Y)$.

<u>Proof</u>. Since X and Y are locally noetherian, we reduce to constructing a similar isomorphism for $G^\cdot \in D^+(Qco(Y))$, where

$$\underline{\underline{R}}f_*: \; D(Qco(X)) \longrightarrow D(Qco(Y))$$

and

$$f^\#: \; D(Qco(Y)) \longrightarrow D(Qco(X)).$$

(We use here [II.7.19] which says that $D^+(Qco(Y)) \longrightarrow D^+_{qc}(Y)$ is an equivalence of categories, and [I.5.6] which says that taking derived functors is compatible with this isomorphism. Note that $\underline{\underline{R}}f_*$ is defined on all of $D(Qco(X))$ since f_* is of finite cohomological dimension on $Qco(X)$.)

In fact, we will construct an isomorphism

$$Trp_f: \; \underline{\underline{R}}f_* f^\#(G^\cdot) \xrightarrow{\sim} G^\cdot$$

for all $G^\cdot \in D(Qco(Y))$.

We apply [I.7.4] to the categories $A = Qco(X)$ and $B = Qco(Y)$, and to the functor $F = f_*$. Now f_* has cohomological dimension n on $Qco(X)$. Let $P \subseteq Ob\, Qco(X)$ be the collection of sheaves of the form L of Lemma 4.1. Then by the two lemmae, P satisfies the hypotheses of loc. cit. and we conclude that $\underline{\underline{L}}(R^n f_*)$ exists, and there is an isomorphism

$$\psi : \underline{\underline{R}}f_* \xrightarrow{\sim} \underline{\underline{L}}(R^n f_*)[-n]$$

of functors from $D(Qco(X))$ to $D(Qco(Y))$.

We apply ψ to $f^{\#}G^{\cdot}$ for $G^{\cdot} \in K(Qco(Y))$, which gives an isomorphism

$$\underline{\underline{R}}f_* f^{\#}(G^{\cdot}) = \underline{\underline{R}}f_*(f^*(G^{\cdot}) \otimes \omega[n])$$

$$\xrightarrow{\sim} \underline{\underline{L}}(R^n f_*)(f^*(G^{\cdot}) \otimes \omega) .$$

Now each sheaf $f^*(G^p) \otimes \omega$ is in P (since $\omega \cong \mathcal{O}_X(-n-1)$), hence is $R^n f_*$-acyclic, so the expression on the right is just

$$R^n f_*(f^*(G^{\cdot}) \otimes \omega).$$

But for each p, the projection formula gives us an isomorphism

$$R^n f_*(f^*(G^p) \otimes \omega) \xrightarrow{\sim} R^n f_*(\omega) \otimes G^p ,$$

and composing with the isomorphism γ of Theorem 3.4, this becomes G^p. Composing all these isomorphisms we have the required isomorphism

$$Trp_f : \underline{\underline{R}}f_* f^{\#}G^{\cdot} \longrightarrow G^{\cdot} .$$

<u>Remarks</u>. 1. Remember that the isomorphism Trp_f we have just defined depends on the isomorphism γ of Theorem 3.4, and so depends apparently on the projective coordinates.

2. If one does not wish to use [II.7.19] and [I.7.4], one can define Trp_f for $G^\cdot \in D^b_{qc}(Y)$ simply by using the projection formula [II.5.6]. The following proposition shows that the two methods of constructing the trace map agree when both are defined.

<u>Proposition 4.4</u>. Let f,X,Y be as in the previous proposition, let $F^\cdot, G^\cdot \in D^b_{qc}(Y)$, and assume that one of $F^\cdot, G^\cdot$ has finite Tor-dimension. Then the following diagram is commutative:

$$\begin{array}{ccc}
F^\cdot \underline{\underline{\otimes}}\, \underline{\underline{R}}f_* f^{\#} G^\cdot & \longrightarrow & \underline{\underline{R}}f_*(f^* F^\cdot \underline{\underline{\otimes}}\, f^{\#} G^\cdot) \\
\downarrow{\scriptstyle \mathrm{Trp}_f} & & \downarrow \\
F^\cdot \underline{\underline{\otimes}}\, G^\cdot & \longleftarrow & \underline{\underline{R}}f_* f^{\#}(F^\cdot \underline{\underline{\otimes}}\, G^\cdot)
\end{array} ,$$

where the upper horizontal arrow is the projection formula [II 5.6] and the right-hand vertical arrow is the isomorphism of Proposition 2.4a.

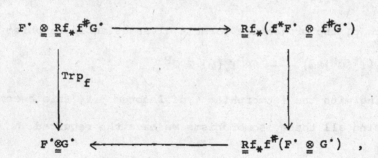

Proof. The question is local on Y, so we may assume Y affine. Then we can take a resolution of $F^{\cdot}$ by direct sums of copies of $\mathcal{O}_Y$. Thus we may work entirely with quasi-coherent sheaves, and will prove the statement for $F^{\cdot}, G^{\cdot} \in D^-(\mathrm{Qco}(Y))$. Then the result follows easily from the definition of the morphisms involved, since if $C^{\cdot\cdot}$ is a Cartan-Eilenberg resolution of $f^{\#}G^{\cdot}$, then $f^*F^{\cdot} \otimes C^{\cdot\cdot}$ is a Cartan-Eilenberg resolution of $f^*F^{\cdot} \otimes f^{\#}G^{\cdot}$.

§5. The duality theorem for projective space.

The duality theorem for projective space now follows easily from what has gone before. At the same time it is a model of how the duality is defined in terms of the functor $f^{\#}$ and the isomorphism Tr_f. When we have a satisfactory functorial theory of $f^!$ (which is $f^{\#}$ in the smooth case) and Tr_f, we will prove the most general duality theorem by reducing to this case (Chapter VII).

Let Y be a noetherian prescheme of finite Krull dimension, let $X = \mathbb{P}^n_Y$, and let $f: X \longrightarrow Y$ be the projection. We define the <u>duality morphism</u>

$$\underline{\theta}_f: \ \underline{\underline{R}}f_* \underline{\underline{R}} \ \underline{\mathrm{Hom}}^{\cdot}_X(F^{\cdot}, f^{\#}G^{\cdot}) \longrightarrow \underline{\underline{R}} \ \underline{\mathrm{Hom}}^{\cdot}_Y(\underline{\underline{R}}f_*F^{\cdot}, G^{\cdot})$$

for $F^{\cdot} \in D^-(X)$ and $G^{\cdot} \in D^+_{qc}(Y)$ by composing the morphism of [II.5.5] with the trace morphism in the second variable (Proposition 4.3).

Applying the functor $\underline{\underline{R}}\Gamma(Y,\cdot)$ to both sides, and using [II.5.2] and [II.5.3] we obtain a global duality morphism

$$\theta_f: \ \underline{\underline{R}} \ \mathrm{Hom}^{\cdot}_X(F^{\cdot}, f^{\#}G^{\cdot}) \longrightarrow \underline{\underline{R}} \ \mathrm{Hom}^{\cdot}_Y(\underline{\underline{R}}f_*F^{\cdot}, G^{\cdot}) \ ,$$

and taking the cohomology of this, we get morphisms

$$\theta_f^i\colon \operatorname{Ext}_X^i(F^\cdot, f^\# G^\cdot) \longrightarrow \operatorname{Ext}_Y^i(\underline{R} f_* F^\cdot, G^\cdot) .$$

Theorem 5.1. Let Y be a noetherian prescheme of finite Krull dimension, and let $X = \mathbb{P}_Y^n$. Then the duality morphisms $\underline{\theta}_f$, θ_f, and θ_f^i are isomorphisms for all $F^\cdot \in D_{qc}^-(X)$ and $G^\cdot \in D_{qc}^+(Y)$.

Proof. Clearly it is sufficient to show that $\underline{\theta}_f$ is an isomorphism. The question is local on Y, so we may assume that Y is affine. Then every quasi-coherent sheaf on Y is a quotient of a direct sum of copies of $\mathcal{O}_Y$, hence, as in Lemma 4.1, every quasi-coherent $\mathcal{O}_X$-module is a quotient of a direct sum of copies of $\mathcal{O}_X(-m)$ for various m. We can take m large, and using any isomorphism $\omega \cong \mathcal{O}_X(-n-1)$ we see that any quasi-coherent sheaf F on X is a quotient of a sheaf of the form

$$L = \oplus\, \omega(-m_i)$$

for certain integers $m_i > 0$.

The functors in question are way-out right in both variables, so by the Lemma on Way-out Functors [I.7.1], we reduce to the case $F^\cdot = L$ of the form above, and $G^\cdot = G$, a single injective quasi-coherent sheaf. Furthermore, $\underline{\underline{R}}\ \underline{Hom}^{\cdot}$ transforms direct sums in the first variable to direct products, so we reduce to the case $F^{\cdot} = \omega(-m)$ for $m > 0$. Thus we have to prove that the map

$$\underline{\theta}_f\colon\ \underline{\underline{R}}f_*\ \underline{\underline{R}}\ \underline{Hom}^{\cdot}_X(\omega(-m),\ f^*(G)\otimes\omega[n])$$

$$\longrightarrow \underline{\underline{R}}\ \underline{Hom}^{\cdot}_Y(\underline{\underline{R}}f_*(\omega(-m)),\ G)$$

is an isomorphism. By [II.5.16] and the projection formula, the complex on the left becomes

$$\underline{\underline{R}}f_*(\mathcal{O}_X(m)) \otimes G[n] = f_*(\mathcal{O}_X(m)) \otimes G[n]$$

since $m > 0$ and $R^if_*(\mathcal{O}_X(m)) = 0$ for $i > 0$ (Theorem 3.4). On the other hand, $R^if_*(\omega(-m)) = 0$ for $i \neq n$, so the complex on the right becomes

$$\underline{Hom}_Y(R^nf_*(\omega(-m)),\ \mathcal{O}_Y) \otimes G[n],$$

using again [II.5.16] and the fact that $R^nf_*(\omega(-m))$ is a locally free sheaf of finite rank on Y.

Now $\underline{\theta}_f$ is the map deduced from the cup-product

$$f_*(\mathcal{O}_X(m)) \times R^nf_*(\omega(-m)) \longrightarrow R^nf_*(\omega)\ ,$$

and so it is an isomorphism by Theorem 3.4. q.e.d.

<u>Corollary 5.2</u>. Let A be a noetherian ring, let $X = \mathbb{P}^n_A$, let I be an A-module, and let $F^\cdot \in D^-_{qc}(X)$. Assume either that I is injective, or that $H^i(X,F^\cdot)$ is projective for all i. Then there is a canonical isomorphism

$$\operatorname{Hom}_A(H^i(X, F^\cdot), I) \cong \operatorname{Ext}^{n-i}_{\mathcal{O}_X}(F^\cdot, \omega \otimes_A I).$$

<u>Remark</u>. When A is a field, $I = A$, and $F^\cdot$ is a complex consisting of a single sheaf, one recovers the duality theorem of Serre for projective space over a field.

§6. Duality for a finite morphism.

Throughout this section we will let $f: X \longrightarrow Y$ be a finite morphism of locally noetherian preschemes. We will define a functor $f^{\flat}$ and a morphism of functors $\mathrm{Trf}_f: \underline{\underline{R}}f_*f^{\flat} \longrightarrow 1$, with the same formal properties as the $f^{\#}$ and Trp_f of §2,4 above. Then we prove a duality theorem similar to the one of §5. This duality theorem is much more elementary than the preceding one, but it is important to set it in the right functorial context.

The reader will notice that the locally noetherian hypothesis is not needed for the definition of $f^{\flat}$, but it is needed for the functorial properties, and the trace map. This suggests that our definition is not the "right" one in the non-noetherian case. On the other hand, we show by an example that the quasi-coherent hypotheses on the sheaves are indeed necessary for a duality theorem.

Let $f: X \longrightarrow Y$ be a finite morphism of locally noetherian preschemes. Let $\bar{f}$ be the morphism of ringed spaces $(X,\mathcal{O}_X) \longrightarrow (Y,f_*\mathcal{O}_X)$, and let $\mathrm{Mod}(f_*\mathcal{O}_X)$ be the category of sheaves of $f_*\mathcal{O}_X$-modules on Y. Then $\bar{f}$ is a flat morphism, and we will consider the functors

$$\overline{f}_*: \operatorname{Mod}(X) \longrightarrow \operatorname{Mod}(f_*\mathcal{O}_X)$$

$$\overline{f}^*: \operatorname{Mod}(f_*\mathcal{O}_X) \longrightarrow \operatorname{Mod}(X).$$

Then $\overline{f}^*$ is exact, since $\overline{f}$ is flat, and the two functors are adjoint [EGA $\underline{0}$, 4.4], i.e., there is a natural map

$$\tau: 1 \longrightarrow \overline{f}_*\overline{f}^*$$

of functors from $\operatorname{Mod}(f_*\mathcal{O}_X)$ into itself, such that the resulting map

$$\operatorname{Hom}_{\mathcal{O}_X}(\overline{f}^*G, F) \longrightarrow \operatorname{Hom}_{f_*\mathcal{O}_X}(G, \overline{f}_*F)$$

is an isomorphism for $F \in \operatorname{Mod}(X)$ and $G \in \operatorname{Mod}(f_*\mathcal{O}_X)$.

<u>Definition</u>. Let $f: X \longrightarrow Y$ be a finite morphism of locally noetherian preschemes. Then we define

$$f^\flat: D^+(Y) \longrightarrow D^+(X)$$

by

$$f^\flat = \overline{f}^* \underline{\underline{R}}\, \underline{\operatorname{Hom}}_{\mathcal{O}_Y}(f_*\mathcal{O}_X, \cdot)\ .$$

(Note that $\underline{\underline{R}}\, \underline{\operatorname{Hom}}_{\mathcal{O}_Y}(f_*\mathcal{O}_X, \cdot)$ is considered as a functor from $D^+(Y)$ to $D^+(\operatorname{Mod}(f_*\mathcal{O}_X))$, and that $\overline{f}^*$ is exact.)

If f has finite Tor-dimension [II §4], then $f_*\mathcal{O}_X$ has finite Tor-dimension in the category Mod(Y), since $f^* = \bar{f}^* \cdot (\otimes f_*\mathcal{O}_X)$. On the other hand, $f_*\mathcal{O}_X$ is coherent, so locally it has a finite resolution by locally free sheaves of finite rank. We conclude that the functor $\underline{\mathrm{Hom}}_{\mathcal{O}_Y}(f_*\mathcal{O}_X,\cdot)$ has finite cohomological dimension, and so in that case we can define

$$f^\flat : D(Y) \longrightarrow D(X)$$

by the same formula as above.

<u>Proposition 6.1</u>. Let $f\colon X \longrightarrow Y$ be a finite morphism of locally noetherian preschemes (resp. with finite Tor-dimension). Then $f^\flat$ takes $D^+_{qc}(Y) \longrightarrow D^+_{qc}(X)$ and $D^+_c(Y) \longrightarrow D^+_c(X)$ (resp. $D_{qc}(Y) \longrightarrow D_{qc}(X)$ and $D_c(Y) \longrightarrow D_c(X)$).

<u>Proof</u>. Follows from [I.7.3],[II.3.2], and the fact that $\bar{f}^*$ takes quasi-coherent sheaves to quasi-coherent sheaves, and coherent sheaves to coherent sheaves.

<u>Proposition 6.2</u>. Let $X \xrightarrow{f} Y \xrightarrow{g} Z$ be two finite morphisms of locally noetherian preschemes (resp. with finite Tor-dimension). Then there is a natural morphism

$$(gf)^\flat \longrightarrow f^\flat g^\flat$$

of functors from $D^+(Z)$ to $D^+(X)$ (resp. $D(Z)$ to $D(X)$). Furthermore, this map is an isomorphism for all $G^\cdot \in D^+_{qc}(Z)$ (resp. $D_{qc}(Z)$).

Proof. For $G \in \mathrm{Mod}(Z)$ there is a natural isomorphism

$$(\overline{gf})^* \underline{\mathrm{Hom}}_{\mathcal{O}_Z}((gf)_* \mathcal{O}_X, G) \xrightarrow{\sim} \overline{f}^* \underline{\mathrm{Hom}}_{\mathcal{O}_Y}(f_* \mathcal{O}_X, \overline{g}^* \underline{\mathrm{Hom}}_{\mathcal{O}_Z}(g_* \mathcal{O}_Y, G)),$$

whence by [I.5.4] the morphism of functors

$$(gf)^\flat \longrightarrow f^\flat g^\flat .$$

To show it is an isomorphism for $G^\cdot \in D^+_{qc}(Z)$ (resp. $D_{qc}(Z)$) we use [I.7.1] and [II.7.18] to reduce to the case where $G^\cdot$ is a single quasi-coherent injective $\mathcal{O}_Z$-module. Then $\overline{g}^* \underline{\mathrm{Hom}}_{\mathcal{O}_Z}(g_* \mathcal{O}_Y, G)$ is a quasi-coherent injective $\mathcal{O}_Y$-module (since for a morphism of rings $A \longrightarrow B$, if I is an injective A-module, then $\mathrm{Hom}_A(B,I)$ is an injective B-module), so we reduce to the isomorphism of sheaves mentioned above, by [II.7.14] and [II.7.16].

Proposition 6.3. Let $f: X \longrightarrow Y$ be a finite morphism of locally noetherian preschemes (resp. with finite Tor-dimension), and let $u: Y' \longrightarrow Y$ be a flat morphism with Y' locally noetherian. Let $X' = X \times_Y Y'$, and let v, g be the projections. Then there is a natural functorial isomorphism

$$\begin{array}{ccc} X' & \xrightarrow{v} & X \\ \downarrow g & & \downarrow f \\ Y' & \xrightarrow{u} & Y \end{array}$$

$$v^* f^\flat (G^\cdot) \xrightarrow{\sim} g^\flat u^*(G^\cdot)$$

for $G^\cdot \in D^+(Y)$ (resp. $G^\cdot \in D(Y)$).

<u>Proof</u>. Use [II.5.8]. Details left to reader.

<u>Corollary 6.4</u>. With the hypotheses of the Proposition, assume furthermore that u (and hence also v) is a smooth morphism. Then there is a natural functorial isomorphism

$$v^\# f^\flat (G^\cdot) \xrightarrow{\sim} g^\flat u^\# (G^\cdot)$$

for $G^\cdot \in D^+(Y)$ (resp. $G^\cdot \in D(Y)$). Moreover, under composition of two such Cartesian diagrams, this isomorphism is compatible with the isomorphisms of Propositions 2.2 and 6.2.

<u>Proof</u>. Follows immediately from the Proposition and the fact that $\omega_{Y'/Y}$ is compatible with arbitrary base extension.

<u>Proposition 6.5</u>. Let $f\colon X \longrightarrow Y$ be a finite morphism of locally noetherian preschemes (resp. with finite Tor-dimension). Then there is a functorial morphism

$$\mathrm{Trf}_f\colon \ \underline{\underline{R}}f_* f^\flat (G^\cdot) \longrightarrow G^\cdot$$

for $G^\cdot \in D^+_{qc}(Y)$ (resp. $G^\cdot \in D_{qc}(Y)$).

<u>Proof</u>. Consider the natural map, for $G \in \mathrm{Mod}(Y)$

$$\tau\colon\ \underline{\mathrm{Hom}}_{\mathcal{O}_Y}(f_*\mathcal{O}_X, G) \longrightarrow \overline{f}_*\overline{f}^* \underline{\mathrm{Hom}}_{\mathcal{O}_Y}(f_*\mathcal{O}_X, G).$$

This gives rise to a functorial morphism [I.5.4]

$$\underline{\underline{R}}\tau\colon\ \underline{\underline{R}}\ \underline{\mathrm{Hom}}_{\mathcal{O}_Y}(f_*\mathcal{O}_X, G^{\cdot}) \longrightarrow \underline{\underline{R}}f_*f^{\flat}(G^{\cdot})$$

for $G^{\cdot} \in D^+(Y)$ (resp. $D(Y)$). I claim $\underline{\underline{R}}\tau$ is an isomorphism for $G^{\cdot} \in D^+_{qc}(Y)$ (resp. $D_{qc}(Y)$). Indeed, using [I.7.3] we reduce to the case where $G^{\cdot} = G$ is a single quasi-coherent injective $\mathcal{O}_X$-module. In that case τ is an isomorphism [EGA II.1.4.3] since f is an affine morphism, and $f^{\flat}(G)$ is injective (as we saw above) so we are done.

Now composing $(\underline{\underline{R}}\tau)^{-1}$ with the natural map

$$\underline{\underline{R}}\ \underline{\mathrm{Hom}}_{\mathcal{O}_Y}(f_*\mathcal{O}_X, G^{\cdot}) \longrightarrow G^{\cdot}$$

derived from the map

$$\underline{\mathrm{Hom}}_{\mathcal{O}_Y}(f_*\mathcal{O}_X, G) \longrightarrow G,$$

"evaluation at one", gives Trf_f.

<u>Proposition 6.6</u>. 1) Let $X \xrightarrow{f} Y \xrightarrow{g} Z$ be a composition of two finite morphisms as in 6.2 above. Then there is a commutative diagram

$$\begin{array}{ccc}
\underline{\underline{R}}(gf)_*(gf)^\flat & \xrightarrow{\operatorname{Trf}_{gf}} & 1 \\
\Big\| \wr\approx & & \Big\downarrow \operatorname{Trf}_g \\
\underline{\underline{R}}g_*\underline{\underline{R}}f_*f^\flat g^\flat & \xrightarrow{\operatorname{Trf}_f} & \underline{\underline{R}}g_*g^\flat
\end{array}$$

of functors on $D^+_{qc}(Z)$ (resp. $D_{qc}(Z)$).

2) Let $u\colon Y' \longrightarrow Y$ be a flat base extension, as in 6.3 above. Then there is a commutative diagram

$$\begin{array}{ccc}
u^*\ \underline{\underline{R}}f_*f^\flat & \xrightarrow{\operatorname{Trf}_f} & u^* \\
\Big\| \approx & & \Big\downarrow \operatorname{Trf}_g \\
\underline{\underline{R}}g_*v^*f^\flat & \xrightarrow{\approx} & \underline{\underline{R}}g_*g^\flat u^*
\end{array}$$

of functors on $D^+_{qc}(Y)$ (resp. $D_{qc}(Y)$). (The left vertical arrow is [II.5.12].)

<u>Proof</u>. Left to reader.

<u>Theorem 6.7</u> (Duality). Let $f\colon X \longrightarrow Y$ be a finite morphism of noetherian preschemes of finite Krull dimension. Then the duality morphism

$$\theta_f\colon\ \underline{\underline{R}}f_*\ \underline{\underline{R}}\ \underline{\operatorname{Hom}}^{\cdot}_X(F^{\cdot},f^\flat G^{\cdot}) \longrightarrow \underline{\underline{R}}\ \underline{\operatorname{Hom}}^{\cdot}_Y(\underline{\underline{R}}f_*F^{\cdot},G^{\cdot})$$

defined by composing [II.5.5] with Trf_f, is an isomorphism for $F^{\cdot} \in D^-_{qc}(X)$ and $G^{\cdot} \in D^+_{qc}(Y)$.

<u>Proof</u>. Making the usual reductions, we arrive at the following well-known statement: let $A \longrightarrow B$ be a homomorphism of rings, let M be a B-module and let N be an A-module. Then the natural map

$$\operatorname{Hom}_B(M, \operatorname{Hom}_A(B,N)) \longrightarrow \operatorname{Hom}_A(M,N)$$

is an isomorphism.

<u>Example</u>. One cannot expect a duality theorem for non-quasi-coherent sheaves, even for a finite étale morphism of integral noetherian schemes. Let Y be a non-singular curve over a field k, algebraically closed, and let X be a double covering of Y. Let $y \in Y$ be a closed point, and let x_1, x_2 be the two points lying over y. Let K(X) and K(Y) be the function fields of X and Y, respectively. Let G be the sheaf K(Y), concentrated at the point y. Then G is a (non-quasi-coherent) indecomposable injective $\mathcal{O}_Y$-module [II.7.11]. One sees easily that $f^\flat(G)$ is the sheaf on X consisting of two copies of K(X), one concentrated at x_1, and one concentrated at x_2. It is the direct sum of two indecomposable injective $\mathcal{O}_X$-modules.

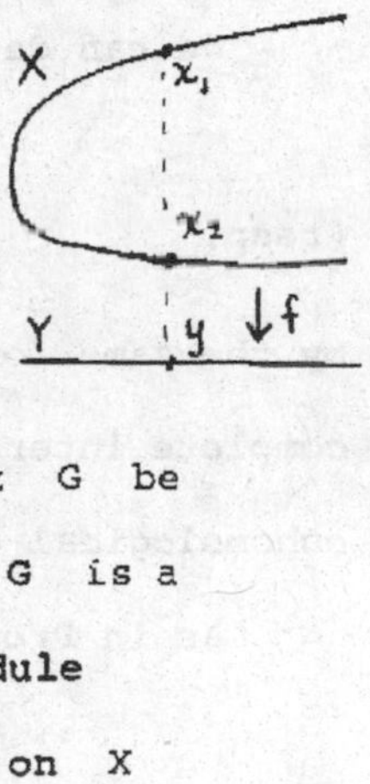

Now let $F = \mathcal{O}_X$. Then we have

$$f_* \underline{\mathrm{Hom}}_{\mathcal{O}_X}(F, f^\flat G) = 2K(X) \text{ concentrated at } y$$

$$\underline{\mathrm{Hom}}_{\mathcal{O}_Y}(f_* F, G) = K(X) \text{ concentrated at } y.$$

Thus the duality morphism $\underline{\theta}_f$ cannot be an isomorphism.

Remark 6.8. Let $f\colon X \longrightarrow Y$ be a closed immersion of preschemes (not necessarily locally noetherian) (resp. with X a locally complete intersection in Y [§1]). Then we can improve on the results of this section as follows.

We can define

$$f^\flat \;:\; D^+(Y) \longrightarrow D^+(X)$$

(resp. $f^\flat \;:\; D(Y) \longrightarrow D(X)$)

by the same formula as above, noting that if X is a local complete intersection, then the functor $\underline{\mathrm{Hom}}_{\mathcal{O}_Y}(f_*\mathcal{O}_X, \cdot)$ has finite cohomological dimension.

As in Proposition 6.2, there is a natural map

$$(gf)^\flat \longrightarrow f^\flat g^\flat$$

which is defined and is an isomorphism on $D^+(Z)$ (resp. $D(Z)$). One need only note that $\overline{f}^*$ and $\overline{g}^*$ are the identity maps, so the reduction to the quasi-coherent case is unnecessary.

The trace map of Proposition 6.4,

$$\mathrm{Trf}_f\colon \quad \underline{\underline{R}}f_* f^{\flat}(G^\cdot) \longrightarrow G^\cdot ,$$

is defined for $G^\cdot \in D^+(Y)$ (resp. $D(Y)$) since the morphism τ of the proof is the identity.

The compatibilities of Proposition 6.6 carry over to this more general case. (Here one needs to note that the quasi-coherence assumption in [II.5.12] is unnecessary if f is a closed immersion, because then f_* is an exact functor on $\mathrm{Mod}(X)$.)

Finally, the duality of Theorem 6.6 is valid for $F^\cdot \in D(X)$ and $G^\cdot \in D^+(Y)$. Indeed, we may assume that $G^\cdot$ is a complex of injective $\mathcal{O}_Y$-modules. Then $f^{\flat}G^\cdot$ is also a complex of injectives; f_* is exact, so we have to show that

$$f_* \, \underline{\mathrm{Hom}}^\cdot_X(F^\cdot, f^{\flat}G^\cdot) \longrightarrow \underline{\mathrm{Hom}}^\cdot_Y(f_*F^\cdot, G^\cdot)$$

is an isomorphism. It is true for each F^p, G^q separately, and hence is true for the complexes. (Note we do not use the Lemma on Way-out Functors this time.)

<u>Proposition 6.9</u>. Let $f: X \longrightarrow Y$ be a finite morphism of locally noetherian preschemes. Then

a) There is a functorial isomorphism

$$f^\flat(F^\cdot) \underline{\underline{\otimes}} \underline{\underline{L}}f^*(G^\cdot) \xrightarrow{\ \sim\ } f^\flat(F^\cdot \underline{\underline{\otimes}} G^\cdot)$$

for $F^\cdot \in D^+(Y)$, $G^\cdot \in D^b(Y)_{fTd}$.

b) There is a functorial isomorphism

$$\underline{\underline{R}}\ \underline{Hom}^\cdot(\underline{\underline{L}}f^*F^\cdot,\ f^\flat G^\cdot) \xrightarrow{\ \sim\ } f^\flat(\underline{\underline{R}}\ \underline{Hom}^\cdot(F^\cdot, G^\cdot))$$

for $F^\cdot \in D_c^-(Y)$ and $G^\cdot \in D_{qc}^+(Y)$.

c) There is a commutative diagram (for $F^\cdot \in D_{qc}^b(Y)$ and $G^\cdot \in D_{qc}^b(Y)_{fTd}$)

$$\begin{array}{ccc} F^\cdot \underline{\underline{\otimes}}\ \underline{\underline{R}}f_* f^\flat G^\cdot & \xrightarrow{[\text{II } 5.6]} & \underline{\underline{R}}f_*(\underline{\underline{L}}f^*F^\cdot \underline{\underline{\otimes}} f^\flat G^\cdot) \\ \Big\downarrow {\scriptstyle Trf_f} & & \Big\downarrow {\scriptstyle a)} \\ F^\cdot \underline{\underline{\otimes}} G^\cdot & \xleftarrow{Trf_f} & \underline{\underline{R}}f_* f^\flat(F^\cdot \underline{\underline{\otimes}} G^\cdot) \end{array}$$

d) There is a commutative diagram (for $F^{\cdot} \in D_c^-(Y)$ and $G^{\cdot} \in D_{qc}^+(Y)$):

$$\begin{array}{ccc}
\underline{\underline{R}}f_*(\underline{\underline{R}}\ \underline{\mathrm{Hom}}_X^{\cdot}(\underline{\underline{L}}f^*F^{\cdot}, f^{\flat}G^{\cdot})) & \xrightarrow{\ \ b)\ \ } & \underline{\underline{R}}f_* f^{\flat}\underline{\underline{R}}\ \underline{\mathrm{Hom}}_Y^{\cdot}(F^{\cdot},G^{\cdot}) \\
\Big\| \ [\mathrm{II}.5.10] & & \Big| \ \mathrm{Trf}_f \\
\underline{\underline{R}}\ \underline{\mathrm{Hom}}_Y^{\cdot}(F^{\cdot},\underline{\underline{R}}f_*f^{\flat}G^{\cdot}) & \xrightarrow{\ \ \mathrm{Trf}_f\ \ } & \underline{\underline{R}}\ \underline{\mathrm{Hom}}_Y^{\cdot}(F^{\cdot},G^{\cdot})
\end{array} .$$

Proof. Left to reader.

§7. The fundamental local isomorphism.

We have seen two different contexts in which we could define a functor $f^!$ giving rise to a duality theorem: the case of a finite morphism and the case of a projective space morphism. (We called them $f^\flat$ and $f^\#$, respectively, to avoid confusion.) In this section we give a local isomorphism which will be the key link relating these two different procedures in the definition of the functor $f^!$ for a general morphism of preschemes.

Let $X = \operatorname{Spec} A$ be an affine scheme, and let $\underline{f} = (f_1,\cdots,f_n)$ be an A-sequence, that is, $f_1,\cdots,f_n$ are elements of A; f_1 is a non-zero divisor in A, and for each $i = 2,\cdots,n$, f_i is a non-zero divisor in the quotient ring $A/(f_1,\cdots,f_{i-1})$. Let J be the sheaf of ideals on X generated by $f_1,\cdots,f_n$, and let F be any sheaf of $\mathcal{O}_X$-modules.

We define the Koszul complex $K_\bullet(\underline{f})$ as follows (cf. [EGA III 1.1] where the sign convention is different):

$$K_p(\underline{f}) = \bigwedge^p (\mathcal{O}_X^n) \qquad p = 0,\cdots,n .$$

If $e_1,\cdots,e_n$ is the usual basis of $\mathcal{O}_X^n$, then

$$d_p\colon K_p(\underline{f}) \longrightarrow K_{p-1}(\underline{f})$$

is defined by

$$d_p(e_{i_1} \wedge \ldots \wedge e_{i_p}) = \sum (-1)^j f_j e_{i_1} \wedge \ldots \wedge \hat{e}_{i_j} \wedge \ldots \wedge e_{i_p} .$$

For any sheaf $F \in \mathrm{Mod}(X)$ we define

$$K^{\cdot}(\underline{f};F) = \underline{\mathrm{Hom}}^{\cdot}_{\mathcal{O}_X}(K_{\cdot}(\underline{f}),F) .$$

Then a section

$$\alpha \in \Gamma(K^p(\underline{f};F))$$

is determined by giving its values

$$\alpha_{i_1,\cdots,i_p} = \alpha(e_{i_1} \wedge \ldots \wedge e_{i_p}) \in \Gamma(X,F) ,$$

and the boundary operator is given by

$$(d\alpha)_{i_1,\cdots,i_{p+1}} = \sum (-1)^j f_j \, \alpha_{i_1,\cdots,\hat{i}_j,\cdots,i_{p+1}} .$$

We denote the cohomology of $K^{\cdot}(\underline{f};F)$ by $H^i(\underline{f};F)$.

Recall [EGA III.1.1] that for $(f_1,\cdots,f_n)$ an A-sequence, $K_{\cdot}(\underline{f})$ is a resolution of $\mathcal{O}_X/J$ by locally free $\mathcal{O}_X$-modules by means of the augmentation $\epsilon\colon K_o(\underline{f}) = \mathcal{O}_X \longrightarrow \mathcal{O}_X/J$. Hence the maps

$$\begin{array}{ccc} & & K^{\cdot}(\underline{f};F) \\ & & \Big| \xi \\ \underline{\underline{R}}\ \underline{\mathrm{Hom}}^{\cdot}_{\mathcal{O}_X}(\mathcal{O}_X/J,F) & \xrightarrow{\epsilon} & \underline{\underline{R}}\ \underline{\mathrm{Hom}}^{\cdot}_{\mathcal{O}_X}(K.(\underline{f}),F) \end{array}$$

are isomorphisms, and we deduce isomorphisms

$$\psi^i\colon \underline{\mathrm{Ext}}^i_X(\mathcal{O}_X/J,F) \xrightarrow{\sim} H^i(\underline{f};F)$$

for $i = 0,\cdots,n$. We define now a map

$$\varphi_{\underline{f}}\colon \ \underline{\mathrm{Ext}}^n(\mathcal{O}_X/J,F) \longrightarrow F/JF$$

by composing ψ^n with the map of $H^n(\underline{f};F) \longrightarrow F/JF$ defined by sending $\alpha \in K^n(\underline{f};F)$ to $\alpha_{1,\cdots,n} \in \Gamma(X,F)$. Then $\varphi_{\underline{f}}$ is an isomorphism. More generally, one shows using the Koszul complex [EGA III 1.1] that there are isomorphisms

(1) $$\underline{\mathrm{Ext}}^i(\mathcal{O}_X/J,F) \cong \underline{\mathrm{Tor}}_{n-i}(\mathcal{O}_X/J,F)$$

for all $i = 0,\cdots,n$. We have made explicit the case $i = n$, noting that $\underline{\mathrm{Tor}}_o(\mathcal{O}_X/J,F) = F \otimes \mathcal{O}_X/J = F/JF$.

Lemma 7.1. Let $X = \mathrm{Spec}\ A$ be an affine scheme; let $\underline{f} = (f_1,\cdots,f_n)$ and $\underline{g} = (g_1,\cdots,g_n)$ be two A-sequences generating the same ideal J, and let $g_i = \sum c_{ij}f_j$ with $c_{ij} \in A$. Let F be a sheaf of $\mathcal{O}_X$-modules. Then there is a commuative diagram

$$\begin{array}{ccc} & \overset{\varphi_{\underline{f}}}{\nearrow} & F/JF \\ \underline{\mathrm{Ext}}^n(\mathcal{O}_X/J,F) & & \downarrow \det(c_{ij}) \\ & \underset{\varphi_{\underline{g}}}{\searrow} & F/JF \end{array}$$

Proof. One has only to note that there is an isomorphism $\wedge \underline{c}$ of $K.(\underline{g})$ into $K.(\underline{f})$ which is given in the pth degree by $\overset{p}{\wedge}(c_{ij})$. In particular its action on the nth degree is $\det(c_{ij})$. The result follows immediately.

Proposition 7.2. (Fundamental Local Isomorphism). Let $i: Y \longrightarrow X$ be a closed immersion of preschemes, where Y is locally a complete intersection in X of codimension n, and let F be a sheaf on X. Then there is a natural functorial isomorphism

$$\varphi: \ \underline{\mathrm{Ext}}^n_{\mathcal{O}_X}(\mathcal{O}_Y,F) \xrightarrow{\ \sim\ } F \otimes_{\mathcal{O}_X} \omega_{Y/X}$$

(cf. §1 for definition of $\omega_{Y/X}$). Furthermore, if F is i^*-acyclic, then

$$\underline{\mathrm{Ext}}^j_{\mathcal{O}_X}(\mathcal{O}_Y, F) = 0 \qquad \text{for } j \neq n .$$

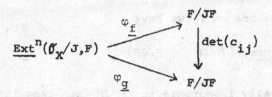

<u>Proof</u>. Let J be the ideal of Y in X. Since $\bigwedge^n J/J^2$ is locally free of rank one on Y, we have

$$F \otimes_{\mathcal{O}_X} \omega_{Y/X} = \underline{\mathrm{Hom}}_{\mathcal{O}_Y}(\bigwedge^n J/J^2, F/JF) ,$$

and this latter is locally isomorphic to F/JF (non-canonically). Thus we can define an isomorphism φ locally by the condition that φ followed by evaluation at $f_1 \wedge \cdots \wedge f_n$ (where $\underline{f} = (f_1, \cdots, f_n)$ is an $\mathcal{O}_X$-sequence generating J locally) be $\varphi_{\underline{f}}$. When one changes basis of J, $\bigwedge^n J/J^2$ changes according to the determinant of the transformation. Therefore by the Lemma we see that the definition of φ is independent of the basis chosen, and hence the local definitions glue together to give a global φ.

To say F is i^*-acyclic is to say that

$$\underline{\mathrm{Tor}}_j^{\mathcal{O}_X}(\mathcal{O}_Y, F) = 0 \qquad \text{for } j \neq 0 .$$

By the isomorphisms (1) above we see that this is equivalent to the condition on the <u>Ext</u>'s of the Proposition.

<u>Corollary 7.3</u>. Let $i\colon Y \longrightarrow X$ and $\omega_{Y/X}$ be as in the proposition. Then there is a natural functorial isomorphism, for all $F^\cdot \in D(X)$,

$$\eta_i\colon\ i^\flat(F^\cdot) \longrightarrow \underline{\underline{L}}i^*(F^\cdot) \otimes \omega_{Y/X}[-n] .$$

Proof. (Note that we write $\otimes$ on the right, not $\underline{\underline{\otimes}}$, because $\omega_{Y/X}$ is locally free on Y and so tensoring by it is an exact functor.) In the first place, if $F^{\cdot}$ is reduced to a single sheaf F, which is i^*-acyclic, then on the left we have the single sheaf $\underline{Ext}^n_{\mathcal{O}_X}(\mathcal{O}_Y, F)$ in degree n, by the Proposition, and on the right we have $F \otimes_{\mathcal{O}_X} \mathcal{O}_Y \otimes_Y \omega_{X/Y}[-n]$ which is isomorphic to it by the isomorphism φ of the Proposition.

In the second place, i^* is a functor of finite cohomological dimension, because its derived functors are the $\underline{Tor}_j^{\mathcal{O}_X}(\mathcal{O}_Y, \cdot)$, and $\mathcal{O}_Y$ locally has a flat resolution of length n, namely the Koszul complex mentioned above. Therefore every $F^{\cdot} \in D(X)$ admits a (left)-resolution by i^*-acyclic $\mathcal{O}_X$-modules.

We are thus in a position to apply [I.7.4]. Let A = Mod(X), B = Mod(Y), let F be the functor $i^*\underline{Hom}_{\mathcal{O}_X}(i_*\mathcal{O}_Y, \cdot)$, and let P be the i^*-acyclic $\mathcal{O}_X$-modules. Then $G = R^nF$ is isomorphic to $i^*(\cdot) \otimes_Y \omega_{Y/X}$ by the Proposition, and every element of P is G-acyclic also by the Proposition. Hence there is a functorial isomorphism

$$\underline{\underline{R}}F \xrightarrow{\sim} \underline{\underline{L}}G[-n]$$

which is just what we want.

<u>Proposition 7.4</u>. a) If $Z \xrightarrow{j} Y \xrightarrow{i} X$ are two closed immersions which are locally complete intersections of codimensions m,n respectively, and if $F^{\cdot} \in D(X)$, then there is a commutative diagram

$$\begin{array}{ccc}
(ij)^{\flat}(F^{\cdot}) & \xrightarrow{\eta_{ij}} & \underline{\underline{L}}(ij)^{*}(F^{\cdot}) \otimes_{Z} \omega_{Z/X}[-m-n] \\
\downarrow \alpha & & \\
j^{\flat} i^{\flat}(F^{\cdot}) & & \downarrow \beta \\
\downarrow \eta_{i} & & \\
j^{\flat}(\underline{\underline{L}}i^{*}(F^{\cdot}) \otimes_{Y} \omega_{Y/X}[-n]) & \xrightarrow{\eta_{j}} & \underline{\underline{L}}j^{*}(\underline{\underline{L}}i^{*}(F^{\cdot}) \otimes_{Y} \omega_{Y/X}[-n]) \otimes_{Z} \omega_{Z/Y}[-m]
\end{array}$$

where α is the isomorphism of Proposition 6.2, and β is obtained by composing the isomorphism $\zeta_{i,j}$ of Definition 1.5 with the isomorphisms of [II 5.4], [II 5.9], and [II 5.13].

b) If $i: Y \longrightarrow X$ is a locally complete intersection of codimension n, and if $f: X' \longrightarrow X$ is a flat morphism, then letting Y' be the fibred product, we have a commutative diagram for $F^{\cdot} \in D(X)$

$$\begin{array}{ccc}
Y' & \xrightarrow{j} & X' \\
\downarrow g & & \downarrow f \\
Y & \xrightarrow{i} & X
\end{array}$$

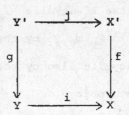

$$\begin{array}{ccc}
g^* i^\flat(F^\cdot) & \xrightarrow{\ \eta_i\ } & g^*[\underline{\underline{L}}i^*(F^\cdot) \otimes_Y \omega_{Y/X}[-n]] \\
\Big\downarrow \alpha & & \Big\downarrow \beta \\
j^\flat f^*(F^\cdot) & \xrightarrow{\ \eta_{i'}\ } & \underline{\underline{L}}j^*(f^*(F^\cdot)) \otimes_{Y'} \omega_{Y'/X'}[-n]
\end{array}$$

where again α and β are composed of the usual identifications.

§8. The functor $f^!$ for embeddable morphisms.

In this section we use the fundamental local isomorphism to relate the functors $f^\#$ and $f^\flat$ defined above, and to define a functor $f^!$ for morphisms which can be factored into a finite morphism followed by a smooth morphism. The main result of this section is only provisional, but it is a model for the stronger results we will obtain in Chapter VII after developing the local techniques.

Lemma 8.1. Let $f: X \to Y$ be a smooth morphism of locally noetherian preschemes, and let $i: Y \to X$ be a section of f. Then there is a functorial isomorphism

$$\psi_{i,f}: G^{\cdot} \xrightarrow{\sim} i^\flat f^\# G^{\cdot}$$

for all $G^{\cdot} \in D(Y)$.

Proof. We first note by Proposition 1.2 that i is a local complete intersection morphism. Hence for any $G^{\cdot} \in D(Y)$ we have

$$i^\flat f^\# G^{\cdot} = i^\flat (f^* G^{\cdot} \otimes \omega_{X/Y}[n])$$

by definition of f, which is isomorphic by the fundamental local isomorphism η_i of Corollary 7.3 to

$$\underline{\underline{L}}i^*(f^*G^\cdot \otimes \omega_{X/Y}[n]) \otimes \omega_{Y/X}[-n].$$

Using [II.5.9] and [II.5.4] this becomes

$$G^\cdot \otimes i^*\omega_{X/Y} \otimes \omega_{Y/X},$$

which finally by the isomorphism $\zeta_{i,f}$ of Definition 1.5 is isomorphic to $G^\cdot$. We compose all these isomorphisms to obtain $\psi_{i,f}$.

Proposition 8.2. (Residue Isomorphism) Let $f\colon X \to Y$ and $g\colon Y \to Z$ be two morphisms of locally noetherian preschemes, with f finite, g smooth, and gf finite. Then there is a functorial isomorphism

$$\psi_{f,g}\colon \quad (gf)^\flat \xrightarrow{\ \sim\ } f^\flat g^\#$$

defined on $D^+_{qc}(Z)$.

Proof. We consider the fibred product $X \times_Z Y$, with projections p_1 and p_2, and let i be the graph morphism of f. Then p_1 is smooth by base extension from g, so i and p_1 satisfy the hypotheses of the Lemma. Thus

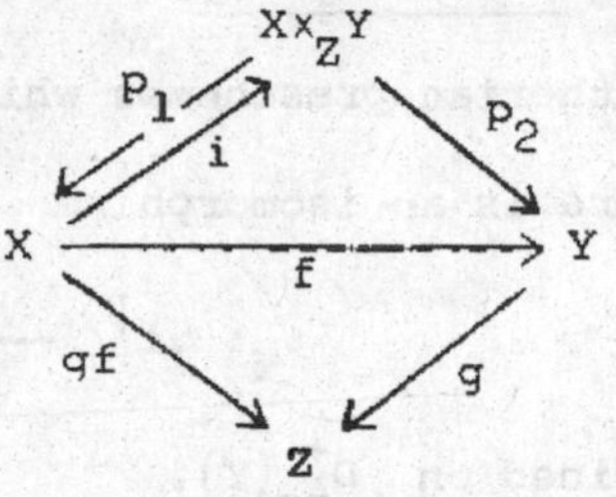

$$(gf)^\flat \xrightarrow{\sim} i^\flat p_1^{\#}(gf)^\flat$$

by ψ_{i,p_1} of the lemma. This in turn is isomorphic to

$$i^\flat p_2^\flat g^{\#}$$

by Corollary 6.4, which is isomorphic finally to

$$f^\flat g^{\#}$$

by Proposition 6.2.

Remarks. This isomorphism, in the case where f is a closed immersion, was first discovered by Grothendieck using a much more complicated procedure. The present proof is due to Cartier, as interpreted by Mumford. This isomorphism will be used in defining the trace map for residual complexes in Chapter VI, an important preliminary to the residue theorem.

Corollary 8.3. Let $f\colon X \longrightarrow Y$ be a morphism of locally noetherian preschemes which is both finite and smooth. Then there is an isomorphism

$$\psi_f\colon f^\flat \xrightarrow{\sim} f^{\#}$$

defined on $D_{qc}^+(Y)$.

Proof. Let f be the identity in the proposition.

Remark. We will leave to the reader the verification that this map is the same as the one deduced from the classical trace map $f_* \mathcal{O}_X \longrightarrow \mathcal{O}_Y$.

Proposition 8.4. Let $f\colon X \to Y$ and $g\colon Y \to Z$ be two morphisms of locally noetherian preschemes, with f finite, g smooth, and gf smooth. Then there is a functorial isomorphism

$$\psi_{f,g}\colon (gf)^{\#} \xrightarrow{\ \sim\ } f^{\flat} g^{\#}$$

defined on $D^+_{qc}(Z)$.

Proof. Considering $X \times_Z Y$ and using the notation of the proof of Proposition 8.2, we have

$$(gf)^{\#} \xrightarrow{\sim} i^{\flat} p_1^{\#}(gf)^{\#} \xrightarrow{\sim} i^{\flat} p_2^{\#} g^{\#} \xrightarrow{\sim} f^{\flat} g^{\#} ,$$

where the isomorphisms are those of Lemma 8.1, Proposition 2.2 (twice), and Proposition 8.2, respectively.

Corollary 8.5. With the same hypotheses as the Proposition, there is a natural map

$$\mathrm{Tr}_f\colon f_* \omega_{X/Z} \longrightarrow \omega_{Y/Z} .$$

Proof. Apply the isomorphism of the Proposition to $\mathcal{O}_Z$, and use the trace map of Proposition 6.5.

<u>Remark</u>. In case Z is the spectrum of a field, X,Y irreducible, and $K(X)/K(Y)$ a separable extension, this trace map coincides with the classical one [3 , Ch. VI §2]. It has the obvious functorial properties: compatibility with composition and flat base extension. It is a non-trivial map, and deserves to be studied more closely.

<u>Proposition 8.6</u>. a) The isomorphisms $\psi_{f,g}$ of Propositions 8.2 and 8.4 are compatible with the isomorphisms of Propositions 2.1 and 6.3 under a flat base extension.

b) If $X \xrightarrow{f} Y \xrightarrow{g} Z \xrightarrow{h} W$ are three morphisms, and if each pair (f,g), (f,hg), (gf,h), (g,h) satisfies the hypotheses of one of Propositions 2.2, 6.2, 8.2, or 8.4, then there is a commutative diagram of the corresponding isomorphisms.

c) If

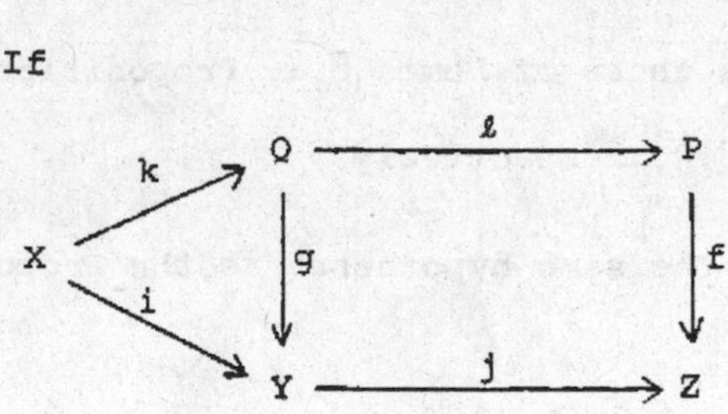

are morphisms with $Q = P \times_Z Y$, f smooth, j,k, and i finite, then there is a commutative diagram

$$\begin{array}{ccccc} (ji)^\flat & \longrightarrow & (\ell k)^\flat f^\# & & \\ \downarrow & & & \searrow & \\ i^\flat j^\flat & \longrightarrow & k^\flat g^\# j^\flat & \longrightarrow & k^\flat \ell^\flat f^\# \end{array}$$

using the isomorphisms of Propositions 6.2, 6.4, and 8.2.

Proof. Left to the patient reader.

Definition. Let S be a fixed prescheme. We say a morphism $f: X \longrightarrow Y$ in the category of preschemes over S is embeddable (or S-embeddable), if there exists a smooth prescheme P over S and a finite morphism $i: X \longrightarrow P_Y = P \times_S Y$ such that $f = p_2 i$. Unless otherwise specified, embeddable will usually mean over $\operatorname{Spec} \mathbb{Z}$.

Examples. A projective morphism $f: X \rightarrow Y$ where Y is quasi-compact and admits an ample sheaf is embeddable (for any S). Indeed, f can be factored through some $\mathbb{P}_Y^N$ [EGA II 5.5.4 (ii)]. Any finite morphism is embeddable, by taking $P = S$. Any morphism of finite type of affine schemes is embeddable in some affine space. Note that any composition of embeddable morphisms is embeddable (!) and that embeddable morphisms are stable under base extension.

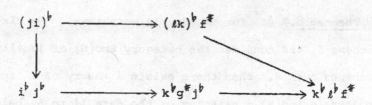

Theorem 8.7 ($f^!$ for embeddable morphisms). We fix a base prescheme S, and consider the category Lno(S) of locally noetherian preschemes over S. Then there exists a theory of $f^!$ for embeddable morphisms in Lno(S) consisting of the data 1) to 5) below, subject to the conditions VAR 1 - VAR 6. Furthermore this theory is unique in the sense that if 1')-5') is another set of such data satisfying VAR 1 - VAR 6, then there is an isomorphism of the functors 1) and 1') compatible with the isomorphisms 2)-5) and 2')-5').

1) For every embeddable morphism $f: X \to Y$ in Lno(S), a functor

$$f^!: D^+_{qc}(Y) \longrightarrow D^+_{qc}(X) .$$

2) For every composition $X \xrightarrow{f} Y \xrightarrow{g} Z$ of embeddable morphisms, an isomorphism of functors

$$c_{f,g}: (gf)^! \longrightarrow f^! g^! .$$

3) For every finite morphism f, an isomorphism

$$d_f: f^! \longrightarrow f^{\flat} .$$

4) For every smooth embeddable morphism f, an isomorphism

$$e_f: f^! \longrightarrow f^{\#} .$$

5) For every embeddable morphism $f: X \longrightarrow Y$, and for every flat base extension $u: Y' \longrightarrow Y$, an isomorphism

$$b_{u,f}: v^*f^! \longrightarrow g^!u^*$$

(where v and g are the two projections of $X' = X \times_Y Y'$).

VAR 1). $c_{f,id} = c_{id,f} = 1$, and there is a commutative diagram of four c's for a composition of three embeddable morphisms.

VAR 2). For a composition of two finite morphisms f,g, compatibility of $c_{f,g}$ with the isomorphism of Proposition 6.2 via d_f and d_g.

VAR 3). Ditto for a composition of smooth morphisms, using Proposition 2.2, e_f and e_g.

VAR 4). For a Cartesian square of embeddable morphisms as in Corollary 6.4, compatibility of that isomorphism with $c_{v,f}$ and $c_{g,u}$ via d_f, d_g, e_u and e_v.

VAR 5). For a composition of two embeddable morphisms f,g satisfying the hypotheses of Proposition 8.2 or 8.4, compatibility of $c_{f,g}$ with $\psi_{f,g}$ via the appropriate d's and e's.

VAR 6). For a flat base extension of a finite or smooth embeddable morphism, compatibility of $b_{u,f}$ with the isomorphism of Proposition 2.1 or 6.3.

Proof. We will give only a sketch, since a similar but more difficult theorem is proved in some detail in Chapter VI.

To define $f^!$ one chooses an f embedding $i: X \longrightarrow P_Y$, and defines $f^! = i^{\flat} p_2^{\#}$. The product of two embeddings is again one, so one shows that $f^!$ is independent of the embedding chosen by using Propositions 8.2 and 8.6b. To define $c_{f,g}$ for a composition, one notes that given embeddings of f and g, say $i: X \longrightarrow P_Y$ and $j: Y \longrightarrow Q_Z$, then

$$(j \times_S P)i: X \longrightarrow (P \times_S Q)_Z$$

is an embedding of gf, and one can define $c_{f,g}$ using the isomorphisms of Corollary 6.4. Of course $c_{f,g}$ is independent of the embeddings chosen ... . One defines d_f and e_f using Propositions 8.2 and 8.4, and $b_{u,f}$ using Proposition 8.6a.

Checking the properties VAR 1 - VAR 6 requires many commutative diagrams, but no imagination. The uniqueness is tedious but straightforward. By the way, the reader will note that 5) and VAR 6 are not needed for the uniqueness statement.

Remarks. One of the main goals of these notes is to obtain a theory of $f^!$, such as the one given in this theorem, for arbitrary morphisms of finite type of locally noetherian preschemes.

The obvious difficulty is that the derived category is not a local object. That is to say, if X is a prescheme, then the presheaf $U \longrightarrow D^+_{qc}(U)$ is not a sheaf of categories on X. One can give a cover of X by open subsets U_i, and complexes $F^\cdot_i \in D^+_{qc}(U_i)$ and isomorphisms $\varphi_{ij} : F^\cdot_i|_{U_{ij}} \longrightarrow F^\cdot_j|_{U_{ij}}$ in $D^+_{qc}(U_{ij})$ which are compatible in $D^+_{qc}(U_{ijk})$, but where there does not exist a complex $F^\cdot \in D^+_{qc}(X)$ whose restriction to U_i is $F^\cdot_i$. Even worse, given two complexes $F^\cdot, G^\cdot \in D^+_{qc}(X)$, and isomorphisms $\varphi_i: F^\cdot|_{U_i} \longrightarrow G^\cdot|_{U_i}$ such that $\varphi_i|_{U_{ij}} = \varphi_j|_{U_{ij}}$ in $D^+_{qc}(U_{ij})$, the φ_i may not glue into a global isomorphism $\varphi: F^\cdot \longrightarrow G^\cdot$.

Thus although every morphism of finite type is <u>locally</u> embeddable, we cannot glue the local functors $f^!$ into a global one.

To overcome this difficulty, we study in Chapter VI the notion of residual complex. These are actual complexes, and hence can be glued. We develop a formalism of $f^!$ for residual complexes similar to the one given here, expanding from the two easy cases of finite and smooth morphisms. Then after proving the duality theorem we can recover a theory of $f^!$ for arbitrary complexes, but only under the additional hypotheses that our schemes be noetherian of finite Krull dimension, and admit a residual complex (e.g., anything of finite type over a regular scheme of finite Krull dimension), and that our complexes have coherent cohomology.

Proposition 8.8. Let $f: X \longrightarrow Y$ be an embeddable morphism of locally noetherian preschemes. Then

6) There is a functorial isomorphism

$$f^!(F^\cdot) \underset{=}{\otimes} \underset{=}{L}f^*(G^\cdot) \xrightarrow{\sim} f^!(F^\cdot \underset{=}{\otimes} G^\cdot)$$

for $F^\cdot \in D^+_{qc}(Y)$ and $G^\cdot \in D^b_{qc}(Y)_{fTd}$.

7) There is a functorial isomorphism

$$\underset{=}{R}\ \underline{Hom}^\cdot(\underset{=}{L}f^*(F^\cdot), f^!(G^\cdot)) \xrightarrow{\sim} f^!(\underset{=}{R}\ \underline{Hom}^\cdot(F^\cdot, G^\cdot))$$

for $F^\cdot \in D^-_c(Y)$ and $G^\cdot \in D^+_{qc}(Y)$.

Proof. Left to reader. (Factor f into a finite morphism followed by a smooth morphism, and use Propositions 2.4 and 6.9.)

Remark. We would like to have an isomorphism such as 6) above when f is flat, $F^\cdot \in D^b_{qc}(Y)_{fTd}$, and $G^\cdot \in D^+_{qc}(Y)$. Both sides make sense in that case, but we do not know how to define a map between them, and hence cannot construct the isomorphism. *However, if Y admits a dualizing complex, we can get a result of this kind for complexes with coherent cohomology [V 8.6].*

§9. The residue symbol.

This section will not be used in the sequel, and so may be omitted at a first reading. In it we define the residue symbol $\mathrm{Res}\left[{\omega \atop t_1,\dots,t_n}\right]$ which is a generalization of the classical notion of residue. For X a non-singular curve over a field k, ω a regular differential form on X, and t a function with an isolated zero at a point P, $\mathrm{Res}\left[{\omega \atop t}\right]$ is just the ordinary residue, at P, of the differential form ω/t. Since we will not use these results later, we leave their proofs to the reader.

Let $f: X \longrightarrow Y$ be a smooth morphism of relative dimension n. Let $t_1,\dots,t_n \in \Gamma(X,\mathcal{O}_X)$ be functions such that the closed subscheme Z of X defined by the ideal $I = (t_1,\dots,t_n)$ is finite and hence flat [EGA IV §11] over Y. Let $\omega \in \Gamma(X,\omega_{X/Y})$ be a global n-differential form on X relative to Y. Under these conditions the residue symbol

$$\mathrm{Res}_{X/Y}\left[{\omega \atop t_1,\dots,t_n}\right] \in \Gamma(Y,\mathcal{O}_Y)$$

can be defined as follows. Let $i: Z \longrightarrow X$ be the inclusion of Z in X, and let $g = fi$. Then by the residue isomorphism $\psi_{i,f}$ of Proposition 8.2 we have an isomorphism

$$g^\flat(\mathcal{O}_Y) \xrightarrow{\ \sim\ } i^\flat f^\#(\mathcal{O}_Y) = i^\flat(\omega_{X/Y}[n])$$

$$\xrightarrow{\ \sim\ } i^*(\omega_{X/Y}) \otimes \omega_{Z/X}$$

$$= \underline{\mathrm{Hom}}_{\mathcal{O}_Z}(\Lambda^n I/I^2, i^*\omega_{X/Y}),$$

where the second arrow is given by the fundamental local isomorphism η_i of Corollary 7.3. (Note that our hypotheses imply that Z is locally a complete intersection in X. Indeed, Z is a local complete intersection in $Z \times_Y X$, by Proposition 1.2, and is defined by $p_2^*(t_1,\cdots,t_n)$, hence this is an $\mathcal{O}_{Z \times_Y X}$-sequence [ZS, vol. II, App. 6, Thm. 2]. But now by faithfully flat descent, $t_1,\cdots,t_n$ is an $\mathcal{O}_X$-sequence.) Now $\Lambda^n I/I^2$ is locally free of rank 1 on Z, so $\bar{t}_1 \wedge \ldots \wedge \bar{t}_n$ is a basis for it. By sending this element into the global section $\bar{\omega}$ of $i^*\omega_{X/Y}$ obtained from ω, we obtain via the isomorphisms above, a global section of $g^\flat(\mathcal{O}_Y)$. Applying g_* we get a global section of

$$g_* g^\flat(\mathcal{O}_Y) = \underline{\underline{R}}\, \underline{\mathrm{Hom}}_{\mathcal{O}_Y}(g_*\mathcal{O}_Z, \mathcal{O}_Y) .$$

But since Z is flat over Y, $g_*\mathcal{O}_Z$ is locally free, and so we can erase the $\underline{\underline{R}}$. Applying our global section to the unit section of $g_*\mathcal{O}_Z$, we obtain a global section of $\mathcal{O}_Y$, which is by definition

$$\mathrm{Res}_{X/Y}\left[\begin{matrix}\omega \\ t_1,\dots,t_n\end{matrix}\right].$$

The residue symbol has the following properties (we assume every time we write a residue symbol that the conditions for its existence are satisfied):

(R0). It is $\mathcal{O}_Y$-linear in ω.

(R1). Let $s_i = \sum c_{ij}t_j$. Then

$$\mathrm{Res}\left[\begin{matrix}\omega \\ t_1,\dots,t_n\end{matrix}\right] = \mathrm{Res}\left[\begin{matrix}\det(c_{ij})\omega \\ s_1,\dots,s_n\end{matrix}\right].$$

In particular, the symbol is alternating in $t_1,\dots,t_n$.

(R2). Localization. It is stable under étale localization on Y.

(R3). Restriction. Let X' be a complete intersection in X, also smooth over Y, defined by functions $s_1,\dots,s_p$ in $\Gamma(X,\mathcal{O}_X)$. Let $t_1,\dots,t_n$ be in $\Gamma(X,\mathcal{O}_X)$, where X is of relative dimension n+p, and let $t_1',\dots,t_n'$ be their restrictions to X'. Let $\omega \in \Gamma(X,\Omega^n_{X/Y})$. Then

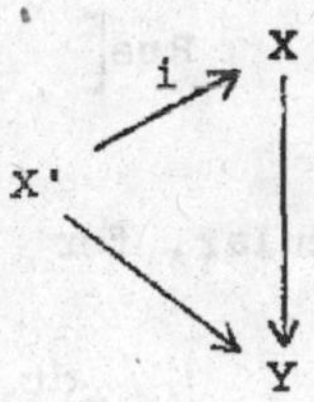

$$\mathrm{Res}_{X'/Y}\left[\begin{matrix}i^*\omega \\ t_1',\dots,t_n'\end{matrix}\right] = \mathrm{Res}_{X/Y}\left[\begin{matrix}\omega \wedge ds_1 \wedge \cdots \wedge ds_p \\ t_1,\dots,t_n,s_1,\dots,s_p\end{matrix}\right].$$

(R4). Transitivity. Let $X \xrightarrow{f} Y \xrightarrow{g} Z$ be two smooth morphisms, of relative dimensions n,p respectively. Let $t_1',\cdots,t_n' \in \Gamma(X,\mathcal{O}_X)$, $\omega' \in \Gamma(X,\omega_{X/Y})$; $s_1,\cdots,s_p \in \Gamma(Y,\mathcal{O}_Y)$, $\omega \in \Gamma(Y,\omega_{Y/Z})$, and let $s_1',\cdots,s_p'$ be the compositions of s_i with f. Then

$$\mathrm{Res}_{X/Z}\begin{bmatrix} \omega\otimes_Y\omega' \\ t_1',\cdots,t_n',s_1',\cdots,s_p' \end{bmatrix} = \mathrm{Res}_{Y/Z}\begin{bmatrix} \omega\cdot\mathrm{Res}_{X/Y}\begin{bmatrix}\omega' \\ t_1',\cdots,t_n'\end{bmatrix} \\ s_1,\cdots,s_p \end{bmatrix} .$$

(R5). Base change. Formation of the residue symbol commutes with base change.

(R6). Trace formula (Normalization). Let $t_1,\cdots,t_n$ and φ be in $\Gamma(X,\mathcal{O}_X)$. Then

$$\mathrm{Res}\begin{bmatrix} \varphi dt_1\wedge\ldots\wedge dt_n \\ t_1,\cdots,t_n \end{bmatrix} = \mathrm{Tr}_{Z/Y}(\varphi|_Z) .$$

In particular, for $\varphi = 1$, one has

$$\mathrm{Res}\begin{bmatrix} dt_1\wedge\ldots\wedge dt_n \\ t_1,\cdots,t_n \end{bmatrix} = \mathrm{rank}\,(Z/Y)\cdot 1_Y .$$

(R7). Intersection formula. For any collection of integers $k_1,\cdots,k_n > 0$, not all equal to one,

$$\mathrm{Res}\begin{bmatrix} dt_1\wedge\ldots\wedge dt_n \\ t_1^{k_1},\cdots,t_n^{k_n} \end{bmatrix} = 0 .$$

(R8). <u>Duality</u>. If $\omega \in \Gamma(\sum t_i \omega_{X/Y})$, then

$$\operatorname{Res}\left[{\omega \atop t_1,\dots,t_n}\right] = 0 ,$$

and conversely if $\operatorname{Res}\left[{f\omega \atop t_1,\dots,t_n}\right] = 0$ for all $f \in \Gamma(X,\mathcal{O}_X)$, then $\omega \in \Gamma(\sum t_i \omega_{X/Y})$.

(R9). <u>Exterior differentiation</u>. For $t_1,\dots,t_n \in \Gamma(X,\mathcal{O}_X)$, and $\omega \in \Gamma(X, \Omega^{n-1}_{X/Y})$ and for $k_1,\dots,k_n > 0$ we have

$$\operatorname{Res}\left[{d\omega \atop t_1^{k_1},\dots,t_n^{k_n}}\right] = \sum k_i \operatorname{Res}\left[{dt_i \wedge \omega \atop t_1^{k_1},\dots,t_i^{k_i+1},\dots,t_n^{k_n}}\right] .$$

(R10). <u>Residue Formula</u>. Let $g\colon X' \longrightarrow X$ be a finite morphism where X',X are both smooth over Y. Let $\omega' \in \Gamma(X',\omega_{X'/Y})$ and let $t_1,\dots,t_n \in \Gamma(X,\mathcal{O}_X)$. Let $t_1',\dots,t_n'$ be their compositions with g. Then

$$\operatorname{Res}_{X'/Y}\left[{\omega' \atop t_1',\dots,t_n'}\right] = \operatorname{Res}_{X/Y}\left[{\operatorname{Tr}_g(\omega') \atop t_1,\dots,t_n}\right] ,$$

where Tr_g is the map of Corollary 8.5.

§10. Trace for projective morphisms.

In this section we show that in the situation of the Residue isomorphism (Proposition 8.2) if g is a projective space morphism, then our trace morphisms Trf for finite morphisms and Trp for projective space morphisms are compatible. This allows us to expand from these two cases to arrive at a theory of the trace map for any projectively embeddable morphism. This result, like the one of §8, is only provisional, because we want eventually a theory of trace for an arbitrary proper morphism. This will come in Chapter VII.

Proposition 10.1. Let Y be a locally noetherian prescheme, let $X = \mathbb{P}^n_Y$, let f be the projection, and let $s: Y \longrightarrow X$ be a section of f. Then for every $G^\cdot \in D^+_{qc}(Y)$, the composition of maps

$$G^\cdot \xrightarrow{\psi_{s,f}} \underline{\underline{R}}f_* \, \underline{\underline{R}}s_* \, s^\flat f^\# G^\cdot \xrightarrow{\mathrm{Trf}_s} \underline{\underline{R}}f_* \, f^\# G^\cdot \xrightarrow{\mathrm{Trp}_f} G^\cdot$$

is the identity. (The maps are those of Lemma 8.1, Proposition 6.5, and Proposition 4.3, respectively.)

Proof. 1) We note that both $\psi_{s,f}$ and Trp_f are calculated by using a Cartan-Eilenberg resolution of $f^\#(G^\cdot) = f^*(G^\cdot) \otimes \omega_{X/Y}[n]$. We can use the same resolution for each, and thus reduce to the case where G is a single quasi-coherent sheaf on Y. Then $f^*(G)\otimes\omega_{X/Y}$ is $\underline{\mathrm{Ext}}^n(s_*\mathcal{O}_Y,\cdot)$-acyclic, and R^nf_*-acyclic, so we have to show that the composition

$$G \longrightarrow f_*\,\underline{\mathrm{Ext}}^n_{\mathcal{O}_X}(s_*\mathcal{O}_Y, f^*G\otimes\omega_{X/Y}) \longrightarrow R^nf_*(f^*G\otimes\omega_{X/Y}) \longrightarrow G$$

is the identity.

2) Noting that the functors above are all right exact in G, and commute with direct sums, and noting that the question is local on Y, we may assume that Y is affine, and thus reduce to the case $G = \mathcal{O}_Y$. Thus we must show that the composition

$$\mathcal{O}_Y \xrightarrow{\ \alpha\ } f_*\,\underline{\mathrm{Ext}}^n_{\mathcal{O}_X}(s_*\mathcal{O}_Y,\omega_{X/Y}) \xrightarrow{\ \beta\ } R^nf_*(\omega_{X/Y}) \xrightarrow{\ \gamma\ } \mathcal{O}_Y$$

is the identity.

3) In other words, from the section s of $\mathbb{P}^n_Y$, we have obtained a map of $\mathcal{O}_Y$ into itself, i.e., a section $\delta(s)\in\Gamma(Y,\mathcal{O}_Y)$, and our problem is to show $\delta(s) = 1$. Since everything in the composition of morphisms in 2) above is flat over Y, this construction is stable under arbitrary base change.

Now our given section s can be obtained from the diagonal section $\Delta\colon \mathbb{P}^n \longrightarrow \mathbb{P}^n\times\mathbb{P}^n$ of projective space over Spec $\mathbb{Z}$ into its product with itself by the base extension $p_2s\colon Y \longrightarrow \mathbb{P}^n$.

Thus $\delta(s) = (p_2 s)^* \delta(\Delta)$ and we reduce to showing that $\delta(\Delta) = 1$. Now $\delta(\Delta)$ is an integer, since $\Gamma(\mathcal{O}_{\mathbb{P}^n}) = \mathbb{Z}$. To find out what integer, it is sufficient to make the base extension at some closed point, say $T_1 = \cdots = T_n = 0$ of $\mathbb{P}^n$, consider $Y = \operatorname{Spec} \mathbb{Z}$, s = the section of $\mathbb{P}^n_Y$ given by $T_1 = \cdots = T_n = 0$, and show that $\delta(s) = 1$ in that case.

4) We show more generally that for any prescheme Y, if s is the section $T_1 = \cdots = T_n = 0$ of $\mathbb{P}^n_Y$, then $\delta(s) = 1$. This is a formidable exercise in explicit calculations, of which we will give a mere outline.

Recalling the notation of §3, we calculate γ (which is the γ of Theorem 3.4) by means of the cover $\mathcal{U} = (U_i)$ of X, the section $\tau = dt_1 \wedge \ldots \wedge dt_n$ of $\omega_{X/Y}|_{U_o}$, and the n-cocycle $\tau / T_o \cdots T_n$ of the Cech complex $f_*(C^{\cdot}(\mathcal{U};\omega_{X/Y}))$.

To calculate α, we use the notation of §7, and the Koszul complex $K^{\cdot}(\underline{t};\omega_{X/Y})$ where $t_1,\cdots,t_n$ are the local coordinates $T_1/T_o,\cdots,T_n/T_o$ on U_o. The map α is obtained by composing the fundamental local isomorphism η_i of Proposition 7.2 with the isomorphism $\zeta_{s,f}$ of Definition 1.5. Recalling that the map $J/J^2 \longrightarrow i^*\Omega^1_{X/S}$ of Proposition 1.2 is defined by sending

$t \in J$ to dt, we see that $\alpha(1)$ is the cocycle

$$\alpha(1) \in K^{\cdot}(\underline{t};\ \omega_{X/Y})$$

given by

$$e_1 \wedge \ldots \wedge e_n \longmapsto \tau\ .$$

Finally, we calculate β by means of the morphism of complexes

$$K^{\cdot}(t_1, \cdots, t_n; \omega) \longrightarrow C^{\cdot}(\mathfrak{U}, \omega)$$

defined by sending a p-cochain

$$\{e_{i_1} \wedge \ldots \wedge e_{i_p} \longmapsto f_{i_1, \cdots, i_p} dt_1 \wedge \cdots \wedge dt_n\}$$

(where $f_{i_1, \cdots, i_p} \in \Gamma(U_o, \mathcal{O}_X)$) to the p-cochain

$$\left\{ \frac{T_o^{p-n-1} f_{i_1, \cdots, i_p} \tau}{T_{i_1} \cdots T_{i_p}} \in \Gamma(U_{o, i_1, \cdots, i_p}, \omega)\ ,\right.$$

and $$\left. 0 \in \Gamma(U_{j_o, j_1, \cdots, j_p}, \omega) \text{ when all } j_i \neq 0 \right\}.$$

Then β applied to the cocycle

$$e_1 \wedge \cdots \wedge e_n \longmapsto dt_1 \wedge \cdots \wedge dt_n$$

gives

$$\tau/T_o \cdots T_n \in \Gamma(U_{o,\ldots,n}, \omega)$$

as required.

<u>Corollary 10.2</u>. The isomorphism γ of Theorem 3.4 is compatible with an automorphism of the projective space, i.e., it is "independent of the choice of homogeneous coordinates".

<u>Proof</u>. Indeed, $\psi_{s,f}$ and Trf_s do not depend on a choice of coordinates, hence Trp_f does not either.

<u>Remark</u>. This shows that for any locally free sheaf E of rank $n+1$ on a prescheme Y, we can define an isomorphism

$$\gamma: \; R^n f_*(\omega) \xrightarrow{\sim} \mathcal{O}_Y$$

where $X = \mathbb{P}(E)$, and $\omega = \omega_{X/Y}$. We can define Trp_f for the projection $f: X \longrightarrow Y$ as in §4, and get a duality theorem as in §5 for this morphism.

<u>Proposition 10.3</u>. Let $u: X \longrightarrow Y$ be a finite morphism of locally noetherian preschemes, let f be the projection of projective n-space over Y, and fill in a Cartesian diagram as shown. Then there is a commutative diagram of morphisms of functors on $D^+_{qc}(Y)$,

$$\begin{array}{ccc} \mathbb{P}^n_X & \xrightarrow{v} & \mathbb{P}^n_Y \\ \downarrow g & & \downarrow f \\ X & \xrightarrow{u} & Y \end{array}$$

$$\begin{array}{ccccc}
\underline{\underline{R}}u_*\underline{\underline{R}}g_*g^{\#}u^{\flat} & \xrightarrow{\ \mathrm{Trp}_g\ } & \underline{\underline{R}}u_*u^{\flat} & & \\
\Big\downarrow{\scriptstyle \alpha} & & & \searrow^{\mathrm{Trf}_u} & \\
\underline{\underline{R}}f_*\underline{\underline{R}}v_*v^{\flat}f^{\#} & \xrightarrow{\ \mathrm{Trf}_v\ } & \underline{\underline{R}}f_*f^{\#} & \xrightarrow{\ \mathrm{Trp}_f\ } & 1\ ,
\end{array}$$

where α is composed of [II.5.1] and Corollary 6.4.

Proof. Left to reader. One follows through the definitions of the maps concerned. The only tricky point is to note that if $C^{\cdot\cdot}$ is a Cartan-Eilenberg resolution of $f^{\#}(G^{\cdot})$, where $G^{\cdot}$ is a complex of quasi-coherent sheaves on Y, then $v^{\flat}(C^{\cdot\cdot})$ is not necessarily a Cartan-Eilenberg resolution of $v^{\flat}f^{\#}(G^{\cdot})$. However, one can find a Cartan-Eilenberg resolution $D^{\cdot\cdot}$ of this latter which dominates it (i.e., there is a map of double complexes $D^{\cdot\cdot} \longrightarrow v^{\flat}(C^{\cdot\cdot})$), which is good enough for the proof.

Proposition 10.4. Let X and Y be locally noetherian preschemes, and let f be a finite morphism of X into $\mathbb{P}^n_Y$. Let $g\colon \mathbb{P}^n_Y \longrightarrow Y$ be the projection, and assume that gf is finite. Then there is a commutative diagram of morphisms of functors on $D^+_{qc}(Y)$,

$$\begin{array}{ccc}
 & & \mathbb{P}^n_Y \\
 & \nearrow^{f} & \Big\downarrow{\scriptstyle g} \\
X & \xrightarrow{\ gf\ } & Y
\end{array}$$

$$\begin{array}{ccc}
\underline{\underline{R}}(gf)_*(gf)^\flat & \xrightarrow{\mathrm{Trf}_{gf}} & 1 \\
\Big\downarrow & & \Big\uparrow \mathrm{Trp}_g \\
\underline{\underline{R}}g_*\underline{\underline{R}}f_*f^\flat g^\# & \xrightarrow{\mathrm{Trf}_f} & \underline{\underline{R}}g_*g^\#
\end{array} ,$$

where the vertical arrow on the left is [II.5.1] composed with the residue isomorphism $\psi_{f,g}$ of Proposition 8.2.

<u>Proof</u>. Considering $\mathbb{P}^n_X = X \times_Y \mathbb{P}^n_Y$ as in the proof of Proposition 8.2, the result follows from Propositions 6.6, 10.1, and 10.3.

<u>Definition</u>. A morphism $f: X \longrightarrow Y$ of preschemes is <u>projectively embeddable</u> if f can be factored $f = pg$ where $p: \mathbb{P}^n_Y \longrightarrow Y$ is the projection of a suitable projective space over Y, and $g: X \longrightarrow \mathbb{P}^n_Y$ is a finite morphism.

<u>Example</u>. If $f: X \longrightarrow Y$ is a projective morphism, where Y is quasi-compact and has an ample sheaf, then f is projectively embeddable [EGA II.5.5.4 (ii)].

<u>Theorem 10.5</u> (Trace for projectively embeddable morphisms). We consider the category (Lno) of locally noetherian preschemes. There is a unique theory of trace for projectively embeddable morphisms $f: X \longrightarrow Y$ in (Lno), consisting of a morphism of

functors $Tr_f: \underline{\underline{R}}f_* f^! \longrightarrow 1$ on $D^+_{qc}(Y)$ for each such morphism f, subject to the conditions TRA 1 - TRA 4 below.

TRA 1). For a composition $X \xrightarrow{f} Y \xrightarrow{g} Z$ of projectively embeddable morphisms, there is a commutative diagram

$$\begin{array}{ccc} \underline{\underline{R}}(gf)_*(gf)^! & \xrightarrow{Tr_{gf}} & 1 \\ \Big\downarrow c_{f,g} & & \Big\uparrow Tr_g \\ \underline{\underline{R}}g_*\underline{\underline{R}}f_*f^!g^! & \xrightarrow{Tr_f} & \underline{\underline{R}}g_*g^! . \end{array}$$

TRA 2). For a finite morphism $f: X \longrightarrow Y$, Tr_f is compatible via d_f with Trf_f.

TRA 3). For the projection $f: \mathbb{P}^n_Y \longrightarrow Y$ of projective space, Tr_f is compatible, via e_f, with Trp_f.

TRA 4). For a projectively embeddable morphism $f: X \longrightarrow Y$ and a flat base extension $u: Y' \longrightarrow Y$, there is a commutative diagram

$$\begin{array}{ccc} u^* \underline{\underline{R}}f_*f^! & \xrightarrow{Tr_f} & u^* \\ \Big\downarrow [II.5.12] & & \Big\uparrow Tr_g \\ \underline{\underline{R}}g_*v^*f^! & \xrightarrow{b_{u,f}} & \underline{\underline{R}}g_*g^!u^* , \end{array}$$

where v and g are the two projections of $X' = X \times_Y Y'$.

<u>Proof</u>. Of course the notations $f^!$, $c_{f,g}$, d_f, e_f, and $b_{u,f}$ refer back to Theorem 8.7.

To construct Tr_f, one chooses a factorization $f = pg$ as in the definition of projectively embeddable morphism above, and defines Tr_f to be the composition of $c_{g,p}$, d_g, e_p, Trf_g, and Trp_p. To show that it is independent of the factorization, one observes that any two factorizations can be dominated by a third, and thus reduces to Proposition 10.4. The properties TRA 1 - TRA 4 are all straightforward, but tedious.

<u>Remarks</u>. The second main object of these notes is to obtain a theory of Tr_f similar to the above one for all <u>proper</u> morphisms f. It is not simply a question of localization, as for the theory of $f^!$, because a proper morphism is not locally projective.

Therefore we resort to an entirely different technique for the construction of the general trace map. We forget entirely the projective case, and work purely from the trace of a finite morphism to define a trace map (which is a map of graded sheaves!) of residual complexes relative to a morphism of finite type. Then we will prove the residue theorem which says that for a proper morphism, the trace map is a morphism of complexes. Finally, after proving the duality theorem, we lift ourselves by our bootstraps, and obtain the general trace map (but under the restrictive hypotheses that our schemes be noetherian of finite Krull dimension admitting a dualizing complex, and that our complexes have coherent cohomology).

§11. Duality for projective morphisms.

Combining the results of §§8 and 10 we have a notion of $f^!$ and Tr_f for projectively embeddable morphisms, and we are in a position to prove the following duality theorem.

Theorem 11.1 (Duality for projectively embeddable morphisms). Let $f: X \longrightarrow Y$ be a projectively embeddable morphism of noetherian preschemes of finite Krull dimension. Then the duality morphism

$$\underline{\theta}_f: \quad \underline{\underline{R}}f_* \underline{\underline{R}}\ \underline{Hom}^{\bullet}_X(F^{\bullet}, f^! G^{\bullet}) \longrightarrow \underline{\underline{R}}\ \underline{Hom}^{\bullet}_Y(\underline{\underline{R}}f_* F^{\bullet}, G^{\bullet}) ,$$

defined by composing [II.5.5] with Tr_f in the second place, is an isomorphism for all $F^{\bullet} \in D^-_{qc}(X)$ and $G^{\bullet} \in D^+_{qc}(Y)$.

Proof. We factor f into pg with g finite and $p: \mathbb{P}^n_Y \longrightarrow Y$ the projection of a suitable projective space. Then, using TRA 1 of Theorem 10.5, and [II §6, ex. 2] we see that $\theta_f = \theta_p \theta_g c_{g,p}$. Thus it is sufficient to show that θ_p and θ_g are isomorphisms. This follows from Theorems 5.1 and 6.7, using the compatibilities of Theorems 8.7 and 10.5.

Remarks. 1. As in §5, the variants θ_f and θ^i_f are also isomorphisms.

2. This result, like the ones of §§8 and 10, is provisional. We will prove a more general duality theorem for proper morphisms in Chapter VII.

3. Taking global sections, and H^o on each side, we have (using [I.6.4])

$$\mathrm{Hom}_{D(X)}(F^{\cdot}, f^{!}G^{\cdot}) \xrightarrow{\sim} \mathrm{Hom}_{D(Y)}(\underline{\underline{R}}f_{*}F^{\cdot}, G^{\cdot}) .$$

For $F^{\cdot} \in D^b_{qc}(X)$ and $G^{\cdot} \in D^b_{qc}(Y)$ this says that $f^!$ (and the map $\mathrm{Tr}_f\colon \underline{\underline{R}}f_{*}f^! \longrightarrow 1$) is a <u>right adjoint</u> of the functor $\underline{\underline{R}}f_{*}$. Therefore the pair $(f^!, \mathrm{Tr}_f)$ is uniquely determined on $D^b_{qc}(Y)$. It is conceivable, however, that there are non-isomorphic functors $f^!$ on $D^+_{qc}(Y)$ each of which gives a duality theorem.

<u>Corollary 11.2</u>. We consider smooth, projectively embeddable morphisms $f\colon X \longrightarrow Y$ of locally noetherian preschemes. Then

a) For each such morphism f of relative dimension n, there is a map

$$\gamma_f\colon\ R^n f_{*}(\omega_{X/Y}) \longrightarrow \mathcal{O}_Y.$$

b) For each pair $f: X \longrightarrow Y$ and $g: Y \longrightarrow Z$ of such morphisms, of relative dimensions n and m, respectively, there is a commutative diagram

$$\begin{array}{ccc}
R^{n+m}(gf)_*(\omega_{X/Z}) & \xrightarrow{\zeta_{f,g}} & R^m g_* R^n f_*(f^*\omega_{Y/Z} \otimes \omega_{X/Y}) \\
 & & \downarrow \approx \\
\downarrow \gamma_{gf} & & R^m g_*(R^n f_*(\omega_{X/Y}) \otimes \omega_{Y/Z}) \\
 & & \downarrow \gamma_f \\
Z & \xleftarrow{\gamma_g} & R^m g_*(\omega_{Y/Z})
\end{array} .$$

(Note that $R^n f_*$ is right exact on quasi-coherent sheaves, so the projection formula gives an isomorphism on the sheaf level.)

c) γ_f commutes with arbitrary base extension. (Note that $R^n f_*$ being right exact, commutes with arbitrary base extension.)

d) For $f: \mathbb{P}^n_Y \longrightarrow Y$, γ_f is the map γ of Theorem 3.4.

e) For $f: X \longrightarrow Y$ a finite smooth morphism,

$$\gamma_f: f_*\mathcal{O}_X \longrightarrow \mathcal{O}_Y$$

is the ordinary classical trace map.

f) For F quasi-coherent on X and G an injective quasi-coherent on Y, the duality map

$$\theta_f^i\colon \operatorname{Ext}^i_{\mathcal{O}_X}(F, f^*G \otimes \omega_{X/Y}) \longrightarrow \operatorname{Hom}_{\mathcal{O}_Y}(R^{n-i}f_*(F), G)$$

defined via γ_f, is an isomorphism.

g) The map γ_f is an isomorphism if and only if f is surjective and has geometrically connected fibres.

<u>Proof</u>. We obtain the map γ_f by applying Tr_f to $G = \mathcal{O}_Y$. Property b) follows immediately from TRA 1.

To prove c), the question is local on Y, so we may assume Y is affine. Then f can be factored $f = pi$, where i is a closed immersion, and $p\colon \mathbf{P}^N_Y \longrightarrow Y$ is a projective space morphism. [EGA II 5.5.4]. We can calculate γ_f by considering the fundamental local isomorphism (Proposition 7.2) which gives

$$\omega_{X/Y} \cong i^* \, \underline{\operatorname{Ext}}^{N-n}_{\mathcal{O}_P}(i_*\mathcal{O}_X, \omega_{P/Y}) \, .$$

Then there is a natural map

$$R^n f_*(\omega_{X/Y}) \longrightarrow R^N p_*(\omega_{P/Y}) \, ,$$

which followed by γ_p gives γ_f. Everything in sight is flat over Y, and hence commutes with arbitrary base extension.

d) follows from TRA 3.

e) we will leave to the reader as an exercise.

f) is a special case of the Theorem, and

g) follows from [EGA III 4.3.1], using f).

Remarks. 1. Later we will prove this theorem for an arbitrary smooth proper morphism of locally noetherian preschemes.

2. In case Y = Spec k with k a field, G = k, X is a smooth projective scheme /k (i.e., "absolutely non-singular projective variety"), and F a coherent sheaf on X, the duality formula f) above reads

$$\operatorname{Ext}^i_X(F, \omega_{X/Y}) \xrightarrow{\sim} H^{n-i}(X;F)^{\vee}$$

where $^{\vee}$ means the dual k-vector space.

CHAPTER IV. LOCAL COHOMOLOGY.

This chapter consists for a great part in definitions, which generalize those of the Local Cohomology lecture notes [LC]. Notable new material is the spectral sequence of a filtered topological space, and the Cousin complex of a sheaf.

§1. Local cohomology groups, sheaves, and complexes.

Throughout this section, X will be an arbitrary topological space, and F a sheaf of abelian groups on X. There are three ways in which one can vary the basic definition of the cohomology of F with supports in a closed subset Z of X: one can replace Z by a family of supports; one can define a relative local cohomology if $Z' \subseteq Z$ are two closed subsets; and one can make everything local by sheafifying. Therefore we will present the definitions in the form of a theme and variations, to allow for all possible combinations of the generalizations suggested above.

We state our results mostly in terms of the cohomology groups, and leave to the reader the appropriate statements in terms of the derived category. When X is an arbitrary topological space, we work in the derived category $D(Ab(X))$ of the category of abelian sheaves on X. If X is a prescheme, we work in the derived category $D(X) = D(Mod(X))$ of the category of $\mathcal{O}_X$-modules. All the derived functors considered in this chapter are compatible in the two cases with the natural functor

$$D^+(X) \longrightarrow D^+(Ab(X)) ,$$

since any injective $\mathcal{O}_X$-module is flasque, and flasque sheaves are acyclic for the functors considered (Motif C below).

Theme.

Let Z be a closed subset of X. Define $\Gamma_Z(X,F)$ to be the group consisting of those global sections of F whose support lies in Z. Define $H^i_Z(X,F)$ to be the ith right derived functor of Γ_Z (which is a left-exact functor), and define $\underline{\underline{R}}\Gamma_Z$ to be the right derived functor on the derived category $D^+(Ab(X))$. These are called the local cohomology groups of F with supports in Z. They have the following properties:

Motif A. If

$$0 \longrightarrow F' \longrightarrow F \longrightarrow F'' \longrightarrow 0$$

is a short exact sequence of abelian sheaves on X, then there is a long exact sequence of groups

$$0 \longrightarrow \Gamma_Z(X,F') \longrightarrow \Gamma_Z(X,F) \longrightarrow \Gamma_Z(X,F") \longrightarrow$$

$$H^1_Z(X,F') \longrightarrow H^1_Z(X,F) \longrightarrow H^1_Z(X,F") \longrightarrow \cdots .$$

<u>Proof</u>. This is the same as to say that Γ_Z is left exact, and so is equal to $H^o(\underline{\underline{R}}\Gamma_Z)$, which it is.

<u>Motif B</u>. There is a long exact sequence

$$0 \longrightarrow \Gamma_Z(X,F) \longrightarrow \Gamma(X,F) \longrightarrow \Gamma(X-Z,F) \longrightarrow$$

$$H^1_Z(X,F) \longrightarrow H^1(X,F) \longrightarrow H^1(X-Z,F) \longrightarrow H^2_Z(X,F) \longrightarrow \cdots .$$

<u>Proof</u>. For any sheaf F one has an exact sequence

$$0 \longrightarrow \Gamma_Z(X,F) \longrightarrow \Gamma(X,F) \longrightarrow \Gamma(X-Z,F) ,$$

and if F is flasque (in particular if F is injective) one can write a zero on the right. Thus if $I^{\bullet}$ is an injective resolution of F, we have an exact sequence of complexes

$$0 \longrightarrow \Gamma_Z(X,I^{\bullet}) \longrightarrow \Gamma(X,I^{\bullet}) \longrightarrow \Gamma(X-Z,I^{\bullet}) \longrightarrow 0$$

which gives rise to a triangle in $D^+(Ab)$,

$$\begin{array}{ccc} & \underline{\underline{R}}\Gamma(X-Z,F) & \\ \swarrow & & \nwarrow \\ \underline{\underline{R}}\Gamma_Z(X,F) & \longrightarrow & \underline{\underline{R}}\Gamma(X,F) \end{array}$$

Taking cohomology gives the result.

<u>Motif C</u>. For F flasque, and $i > 0$, $H_Z^i(X,F) = 0$.

<u>Proof</u>. This follows from Motif B, and the fact that $H^i(X,F) = H^i(X-Z,F) = 0$ for $i > 0$, and $\Gamma(X,F) \longrightarrow \Gamma(X-Z,F)$ is surjective for F flasque.

<u>Remark</u>. In fact, F is flasque $\Longleftrightarrow$ for every closed subspace Z of X, $H_Z^1(X,F) = 0$ [LC, 1.10].

<u>Variation 1</u>.

<u>Definition</u>. A <u>family of supports</u> on a topological space X is a set φ of closed subsets of X such that

(a) if $Z \in \varphi$, and Z' is a closed subset of Z, then $Z' \in \varphi$, and

(b) if $Z_1, Z_2 \in \varphi$, then $Z_1 \cup Z_2 \in \varphi$.

Now let φ be a family of supports, and define $\Gamma_\varphi(X,F)$ to be the group of global sections of F whose support is in φ. Define $\underline{\underline{R}}\Gamma_\varphi(X,F)$ and $H_\varphi^i(X,F)$ to be the right derived functors.

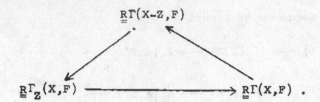

Motif A. Repeat as above, with Z replaced by φ.

Motif D. $H^i_\varphi(X,F) = \varinjlim_{Z\in\varphi} H^i_Z(X,F)$.

Proof. It is true for $i = 0$, and direct limits commute with cohomology of complexes.

Motif C. For F flasque, and $i > 0$, $H^i_\varphi(X,F) = 0$.

Proof. Follows from Motif D, and Motif C above.

Variation 2.

Let $Z' \subseteq Z$ be closed subsets of X. Define $\Gamma_{Z/Z'}(X,F)$ to be $\Gamma_Z(X,F)/\Gamma_{Z'}(X,F)$. Define $\underline{\underline{R}}\Gamma_{Z/Z'}$ and $H^i_{Z/Z'}$ to be the right derived functors. Note that in general $\Gamma_{Z/Z'}$ is not left exact, so that $\Gamma_{Z/Z'} \neq H^0_{Z/Z'}$.

Motif A. Repeat with H^0 in place of Γ, and Z/Z' in place of Z.

Motif B. There is a long exact sequence

$$\begin{aligned}0 \longrightarrow \Gamma_{Z'}(X,F) &\longrightarrow \Gamma_Z(X,F) \longrightarrow H^0_{Z/Z'}(X,F) \longrightarrow \\ H^1_{Z'}(X,F) &\longrightarrow H^1_Z(X,F) \longrightarrow H^1_{Z/Z'}(X,F) \longrightarrow \\ H^2_{Z'}(X,F) &\longrightarrow \cdots .\end{aligned}$$

Motif C. Repeat.

Variation 3.

Let Z be a closed subset of X. Define $\underline{\Gamma}_Z(F)$ to be the sheaf whose sections over an open set U are the elements of $\Gamma_{Z\cap U}(U,F|_U)$. Define $\underline{\underline{R}}\underline{\Gamma}_Z(F)$ and $\underline{H}^i_Z(F)$ to be the right derived functors, which are now complexes of sheaves, resp. sheaves on X.

Motif A. Repeat with underlines.

Motif B. There is a four-term exact sequence

$$0 \longrightarrow \underline{\Gamma}_Z(F) \longrightarrow F \longrightarrow j_*(F|_{X-Z}) \longrightarrow \underline{H}^1_Z(F) \longrightarrow 0$$

where $j: X-Z \longrightarrow X$ is the inclusion, and there are isomorphisms for $i > 0$,

$$R^i j_*(F|_{X-Z}) \xrightarrow{\sim} \underline{H}^{i+1}_Z(F).$$

Motif C. For F flasque and $i > 0$, $\underline{H}^i_Z(F) = 0$.

Proof. One sees easily that for any F and for any i, $\underline{H}^i_Z(F)$ is the sheaf associated to the presheaf

$$U \longrightarrow H^i_{Z\cap U}(U,F|_U) .$$

Now since the restriction of a flasque sheaf to an open subset is flasque, the result follows from Motif C above.

Motif E. There is a spectral sequence

$$E_2^{pq} = H^p(X, \underline{H}_Z^q(F)) \Longrightarrow E^n = H_Z^n(X,F) \ ,$$

or equivalently, in terms of the derived categories,

$\underline{\underline{R}}\Gamma_Z = \underline{\underline{R}}\Gamma \cdot \underline{\underline{R}}\underline{\Gamma}_Z$.

Proof. Referring to [I.5.4], we need only show that $\underline{\Gamma}_Z$ takes injectives into Γ-acyclic objects. Indeed, any injective is flasque, and $\underline{\Gamma}_Z$ of a flasque is flasque, and any flasque is Γ-acyclic.

Variation 4.

Combining variations 1 and 2, let $\psi \subseteq \varphi$ be two families of supports. Define $\Gamma_{\varphi/\psi}(X,F)$ and its derived functors $\underline{\underline{R}}\Gamma_{\varphi/\psi}$ and $H^i_{\varphi/\psi}$.

Motif A. Repeat.

Motif B. Repeat.

Motif C. Repeat.

Motif D. $H^i_{\varphi/\psi}(X,F) = \varinjlim_{\substack{Z \in \varphi \\ Z' \in \psi \\ Z' \subseteq Z}} H^i_{Z/Z'}(X,F)$.

Variation 5.

Combining variations 1 and 3 leads us to a

Definition. A sheaf of families of supports on a topological space X is a sheaf of sets $\underline{\varphi}$, such that for every open set U, $\underline{\varphi}(U)$ is a family of supports on U, and such that for $V \subset U$ the restriction map $\underline{\varphi}(U) \longrightarrow \underline{\varphi}(V)$ is given by $Z \in \underline{\varphi}(U)$ goes to $Z \cap V \in \underline{\varphi}(V)$.

Remarks. 1. If φ is a family of supports on X, we can define a sheaf of families of supports $\tilde{\varphi}$ on X by taking the sheaf associated to the presheaf

$$U \longrightarrow \{Z \cap U \mid Z \in \varphi\}.$$

2. If $\underline{\varphi}$ is a sheaf of families of supports on X, we can define a family of supports $\Gamma(\underline{\varphi})$ on X by $\Gamma(\underline{\varphi}) = \underline{\varphi}(X)$.

3. Note that in general, the operations Γ and $\sim$ are not inverses to each other. For example, if φ is the family of compact subsets of a locally compact Hausdorff space, then $\tilde{\varphi}$ is the maximal sheaf of families of supports, and $\Gamma(\tilde{\varphi})$ is the maximal family of supports. But if X is not compact, $\varphi \neq \Gamma(\tilde{\varphi})$.

4. Conversely, let $f: X \longrightarrow Y$ be a morphism of finite type of locally noetherian preschemes. Let p be an integer, and let $\underline{\varphi}$ be the sheaf of families of supports on X given by $\underline{\varphi}(U)$ = the set of (relatively) closed subsets Z of U such that for every $y \in f(U)$, $Z \cap X_y$ is of codimension $\geq p$ in the fibre X_y. Then in general $\Gamma(\underline{\varphi})^{\sim}$ is different from $\underline{\varphi}$. (For a specific example, let $Y = \operatorname{Spec} k[x,y]$, let $X = \mathbb{P}^1_Y$ and let $p = 1$. Then if $Z \subseteq X$ is obtained by blowing up the origin of Y, Z is locally in $\underline{\varphi}$ except over the origin of Y, but $Z \notin \Gamma(\underline{\varphi})$.)

5. For another example of a family of supports, let Z be a subset of the topological space X, stable under specialization (e.g., $Z = \{x \in X$ of codimension $\geq p\}$ for some p. By the <u>codimension</u> of a point x in a topological space X, we mean the largest integer n, or $+\infty$, such that there exists a sequence $x_0, x_1, x_2, \cdots, x_n = x$ of points of X where for each i, $x_i \rightarrow x_{i+1}$ is a <u>proper specialization</u>, i.e., $x_{i+1} \in \overline{\{x_i\}}$ and $x_i \neq x_{i+1}$). Then the set of subsets of finite unions of closures of points of Z is a family of supports φ. One can also consider the associated sheaf of families of supports $\widetilde{\varphi}$. If X is locally noetherian (as a topological space) and if every closed irreducible

subset of X has a unique generic point, then there is a 1-1 correspondence between subsets of X, stable under specialization, and families of supports φ, such that $\varphi = \Gamma(\tilde{\varphi})$. In these cases we will write Γ_Z for Γ_φ, H_Z^i for H_φ^i, $\underline{\Gamma}_Z$ for $\underline{\Gamma}_{\tilde{\varphi}}$, etc.

Now let $\underline{\varphi}$ be a sheaf of families of supports on X. Define $\underline{\Gamma}_{\underline{\varphi}}(F)$ to be the sheaf whose sections on an open set U are $\Gamma_{\underline{\varphi}(U)}(U,F|_U)$. Define $\underline{\underline{R}}\underline{\Gamma}_{\underline{\varphi}}(F)$ and $\underline{H}^i_{\underline{\varphi}}(F)$ to be the right derived functors.

<u>Motif A</u>. Repeat.

<u>Motif C</u>. Repeat.

<u>Motif D</u>. If $\underline{\varphi}$ is a sheaf of families of supports <u>of global nature</u> (i.e., such that $\underline{\varphi} = \Gamma(\underline{\varphi})^{\sim}$), then

$$\underline{H}^i_{\underline{\varphi}}(F) = \varinjlim_{Z \in \underline{\varphi}(X)} \underline{H}^i_Z(F) .$$

<u>Motif E</u>. There is a spectral sequence

$$E_2^{pq} = H^p(X, \underline{H}^q_{\underline{\varphi}}(F)) \Longrightarrow E^n = H^n_{\Gamma(\underline{\varphi})}(X,F) ,$$

or, in terms of the derived categories, $\underline{\underline{R}}\Gamma_{\Gamma(\underline{\varphi})} = \underline{\underline{R}}\Gamma \cdot \underline{\underline{R}}\underline{\Gamma}_{\underline{\varphi}}$.

Variation 6.

Combining variations 2 and 3, let $Z' \subseteq Z$ be closed subsets of X. Define $\underline{\Gamma}_{Z/Z'}$, $\underline{\underline{R\Gamma}}_{Z/Z'}$, and $\underline{H}^i_{Z/Z'}$. Repeat Motifs A, B, C, and E.

Variation 7.

Combining variations 1, 2, and 3, let $\underline{\psi} \subseteq \underline{\varphi}$ be two sheaves of families of supports on X. Define $\underline{\Gamma}_{\underline{\varphi}/\underline{\psi}}$, $\underline{\underline{R\Gamma}}_{\underline{\varphi}/\underline{\psi}}$, and $\underline{H}^i_{\underline{\varphi}/\underline{\psi}}$. Repeat Motifs A, B, C, D, and E.

Variation 8.

In this case we define a purely punctual invariant. Let x be a point of X and define $\Gamma_x(F)$ to be the subgroup of the stalk F_x consisting of elements $\bar{s}$ which have a representative s in a suitable neighborhood U of x, whose support is $\{\bar{x}\} \cap U$. Define the right derived functors $\underline{\underline{R}}\Gamma_x(F)$ and $H^i_x(F)$. Note that $H^i_x(F) = \underline{H}^i_Z(F)_x$, where $Z = \{\bar{x}\}$, and the subscript x denotes the stalk. Repeat Motifs A and C.

Motif F. Assume that X is a locally noetherian topological space and that every closed irreducible subset of X has a unique generic point. Let $Z' \subseteq Z$ be two subsets of X, stable under specialization, and such that every $x \in Z-Z'$ is maximal in Z (i.e., if $x \in Z$ and $x \longrightarrow x'$ is a non-trivial specialization, then $x' \in Z'$). Let $F^{\cdot} \in D^{+}(Ab(X))$ be a complex of sheaves. Then there is a canonical functorial isomorphism

$$\underline{H}^{i}_{Z/Z'}(F^{\cdot}) \xrightarrow{\sim} \coprod_{x \in Z-Z'} i_x(H^i_x(F^{\cdot})) ,$$

where for any group G, $i_x(G)$ is the constant sheaf G on $\overline{\{x\}}$, and 0 elsewhere. (By abuse of notation we write Z,Z' instead of the sheaves of families of supports they define as in Remark 5 of Variation 5 above.)

Proof. Since both sides are derived functors, it will be sufficient to establish a canonical functorial isomorphism for a single sheaf $F \in Ab(X)$

$$\underline{\Gamma}_{Z/Z'}(F) \xrightarrow{\sim} \coprod_{x \in Z-Z'} i_x(\Gamma_x(F)) .$$

For an open set U we define a map

$$\Gamma_{Z \cap U}(U,F|_U) \longrightarrow \coprod_{x \in (Z-Z') \cap U} \Gamma_x(F)$$

by sending a section into its germ at each stalk. Only finitely many are non-zero, because the support of any section s is a finite union $\overline{\{z_1\}} \cup \cdots \cup \overline{\{z_n\}}$ with $z_i \in Z$

(since X is locally noetherian!) and because each $x \in Z-Z'$ is maximal in Z. A section s goes to zero if and only if it is in $\Gamma_{Z'\cap U}(U, F|_U)$. Thus we have defined an inclusion of sheaves above. Finally, we see that it is surjective, because every germ $\bar{s} \in \Gamma_x(F)$ comes from a section s of F in a suitable neighborhood U, with support $\{\bar{x}\} \cap U$.

<u>Coda</u>.

Having made all these generalizations of the notion of local cohomology, we are practically back where we started:

<u>Motif G</u>. (<u>The Spectral Sequence of a Filtered Topological Space</u>). Let

$$X = \underline{\varphi}^0 \supseteq \underline{\varphi}^1 \supseteq \cdots$$

be a filtration of X by sheaves of families of supports, and let $F^{\cdot}$ be a complex of abelian sheaves on X bounded below. Then there is a spectral sequence

$$E_1^{pq} = \underline{H}^{p+q}_{\underline{\varphi}^p/\underline{\varphi}^{p+1}}(F^{\cdot}) \Longrightarrow E^n = H^n(F^{\cdot}) ,$$

which is biregular [EGA 0_{III} §11] if the filtration is <u>finite</u>, i.e., $\underline{\varphi}^n = \emptyset$ for some n. Or, in terms of the derived category, there is a diagram of triangles (shown in the convergent case)

$$\begin{array}{ccccccccc}
& \underline{\underline{R}}\Gamma_{\underline{\Phi}^{n-1}/\underline{\Phi}^{n}}(F^\cdot) & & & \cdots & & \underline{\underline{R}}\Gamma_{\underline{\Phi}^{0}/\underline{\Phi}^{1}}(F^\cdot) & & \\
\overset{\cdot}{\swarrow} & & \nwarrow & & & \overset{\cdot}{\swarrow} & & \nwarrow & \\
0=\underline{\underline{R}}\Gamma_{\underline{\Phi}^{n}}(F^\cdot) & \longrightarrow & \underline{\underline{R}}\Gamma_{\underline{\Phi}^{n-1}}(F^\cdot) & \longrightarrow \cdots \longrightarrow & & \underline{\underline{R}}\Gamma_{\underline{\Phi}^{1}}(F^\cdot) & \longrightarrow & \underline{\underline{R}}\Gamma_{\underline{\Phi}^{0}}(F^\cdot)=F^\cdot . &
\end{array}$$

Proof. This is just the spectral sequence of a filtered complex [M, Ch. XV]. Take an injective resolution $I^\cdot$ of $F^\cdot$. Then $I^\cdot$ is filtered

$$I^\cdot = \underline{\Gamma}_{\underline{\Phi}^{0}}(I^\cdot) \supseteq \underline{\Gamma}_{\underline{\Phi}^{1}}(I^\cdot) \supseteq \cdots \supseteq \Gamma_{\underline{\Phi}^{n}}(I^\cdot) = 0 ,$$

and the quotients are $\underline{\Gamma}_{\underline{\Phi}^{i}/\underline{\Phi}^{i+1}}(I^\cdot)$.

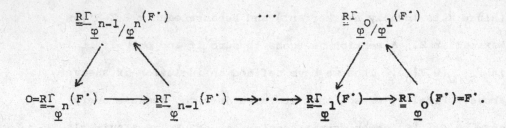

§2. Depth and the Cousin Complex.

Throughout this section, X will denote a locally noetherian topological space in which every closed irreducible subset has a unique generic point.

Definition. Let $F^{\cdot}$ be a complex of sheaves of abelian groups (bounded below) on X and $\underline{\varphi}$ a sheaf of families of supports. Then the $\underline{\varphi}$-depth of $F^{\cdot}$ is the largest integer n (or $+\infty$) such that $\underline{H}^{i}_{\underline{\varphi}}(F^{\cdot}) = 0$ for all $i < n$.

Remark. If X is a locally noetherian prescheme, $\underline{\varphi}$ the family of subsets of a closed subset Z of X, and F a coherent sheaf, then this definition of depth coincides with the usual definition of the Z-depth of F [LC.3.8].

Proposition 2.1. Let $Z' \subseteq Z$ be subsets of (the locally noetherian topological space) X, stable under specialization, and such that every $x \in Z-Z'$ is maximal (with respect to specialization) in Z. Let F be an abelian sheaf on X. Then the following conditions are equivalent:

(i) The natural maps

$$F \longleftarrow \underline{\Gamma}_Z(F) \longrightarrow \underline{H}^0_{Z/Z'}(F)$$

are isomorphisms.

(ii) There is an isomorphism

$$F \cong \coprod_{x \in Z-Z'} i_x(M_x)$$

for suitable choice of abelian groups M_x. (Recall that for any abelian group M, $i_x(M)$ is the constant sheaf M on $\{\overline{x}\}$, and 0 elsewhere.)

(iii) F has supports in Z, Z'-depth ≥ 1, and is flasque.

(iv) F has supports in Z, and Z'-depth ≥ 2.

<u>Proof</u>. (i) $\Longrightarrow$ (ii) follows immediately from Motif F of Variation 8 above.

(ii) $\Longrightarrow$ (iii) Condition (ii) implies that F has supports in Z, and that $\underline{\Gamma}_{Z'}(F) = 0$, i.e., Z'-depth $F \geq 1$. F is flasque because it is a direct sum of sheaves $i_x(M)$, each of which is a constant sheaf on an irreducible space, hence flasque.

(iii) $\Longrightarrow$ (iv) Since F is flasque, $\underline{H}^i_{Z'}(F) = 0$ for all $i > 0$, by Motif C. But $\underline{H}^0_{Z'}(F) = 0$ since F has Z'-depth ≥ 1, so F has Z'-depth $+\infty \geq 2$.

(iv) $\Longrightarrow$ (i) $\underline{\Gamma}_Z(F) \longrightarrow F$ is an isomorphism since F has supports in Z. Now by Motif B there is an exact sequence

$$0 \longrightarrow \underline{\Gamma}_{Z'}(F) \longrightarrow \underline{\Gamma}_Z(F) \longrightarrow \underline{H}^0_{Z/Z'}(F) \longrightarrow \underline{H}^1_{Z'}(F) \longrightarrow \cdots .$$

Since Z'-depth F $\geq$ 2, the two outside terms are zero, so the middle is an isomorphism.

Definition. If F satisfies the equivalent conditions above, we say F lies on the Z/Z'-skeleton of X.

We now come to the Cousin complex, but prove a lemma first.

Lemma 2.2. Let Z' $\subseteq$ Z be as in the previous proposition, and let F be a sheaf on X with supports in Z. Then there is a unique (up to unique isomorphism) Z'-isomorphism

$$\alpha: F \longrightarrow G$$

of F into a sheaf G which lies on the Z/Z'-skeleton. (By Z'-isomorphism we mean a homomorphism of sheaves whose kernel and cokernel have supports in Z'.)

Proof. (1) To show the existence of α, we take $G = \underline{H}^{0}_{Z/Z'}(F)$, and let α be the natural map of $F = \underline{\Gamma}_{Z}(F)$ into G. Then from the exact sequence of Motif B we see that α is a Z'-isomorphism, and from Motif F we see that G lies on the Z/Z'-skeleton.

(2) To see that α is unique up to isomorphism, let $\beta: F \longrightarrow G'$ be another such Z'-isomorphism. Applying the functor $\underline{H}^{0}_{Z/Z'}$, (which takes Z'-isomorphisms into isomorphisms) we get a commutative diagram

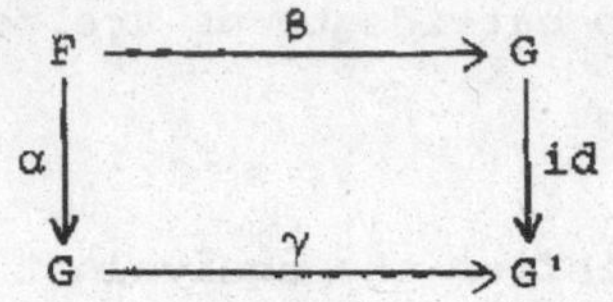

where $\gamma = \underline{H}^{0}_{Z/Z'}(\beta)$ is an isomorphism.

(3) To show that the isomorphism in γ is unique, let $\delta\colon G \longrightarrow G'$ be any other homomorphism which gives a commutative diagram

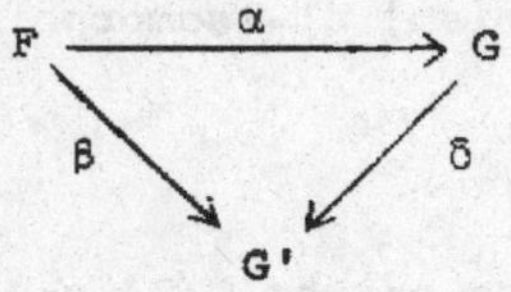

.

Then applying the functor $\underline{H}^{0}_{Z/Z'}$, we obtain a commutative diagram

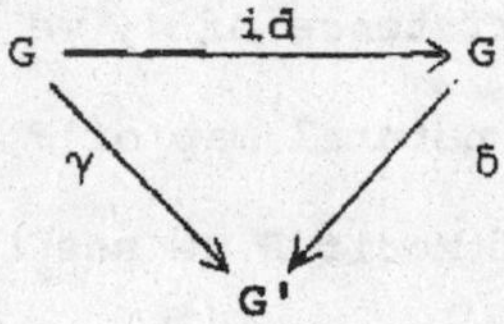

.

<u>Proposition 2.3</u>. Let $X = Z^0 \supseteq Z^1 \supseteq \cdots$ be a filtration of X by subsets Z^p stable under specialization, and such that for each p, each $x \in Z^p - Z^{p+1}$ is maximal in Z^p. Let F be an abelian sheaf on X. Then there is a unique (up to unique isomorphism of complexes) augmented complex

$$\begin{array}{ccccccc} & & F & & & & \\ & & \downarrow & & & & \\ 0 & \longrightarrow & C^0 & \longrightarrow & C^1 & \longrightarrow & \cdots \end{array}$$

with the following properties:

(a) For each $p \geq 0$, C^p lies on the Z^p/Z^{p+1}-skeleton.

(b) For each $p > 0$, $H^p(C^\cdot)$ has supports in Z^{p+2}.

(c) The map $F \longrightarrow H^0(C^\cdot)$ has kernel with supports in Z^1, and cokernel with supports in Z^2.

Furthermore, $C^\cdot$ depends functorially on F.

Proof. We prove by induction on p, that there exists a unique (up to unique isomorphism of complexes) augmented complex

$$\begin{array}{ccccccccc} & & F & & & & & & \\ & & \downarrow & & & & & & \\ 0 & \longrightarrow & C^0 & \longrightarrow & C^1 & \longrightarrow & \cdots & \longrightarrow & C^p \end{array}$$

with all the properties above, except that instead of saying $H^p(C^\cdot)$ has supports in Z^{p+2}, we say that $C^p/\text{Im } C^{p-1}$ has supports in Z^{p+1} (or, if $p = 0$, instead of saying that the cokernel of $F \longrightarrow H^0(C^\cdot)$ has supports in Z^2, we say that $C^0/\text{Im } F$ has supports in Z^1).

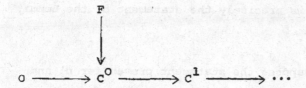

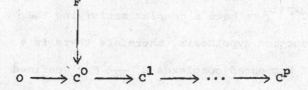

Case p = 0. This is precisely the statement of the Lemma, with $Z = Z^0$ and $Z' = Z^1$.

Induction Step. Suppose the statement proven for p, and take a complex $C^{\cdot}$ as above, defined in degrees $\leq p$. Apply the Lemma to $C^p/\mathrm{Im}\ C^{p-1}$ with $Z = Z^{p+1}$ and $Z' = Z^{p+2}$, and let C^{p+1} be the sheaf G thus obtained. Now since

$$C^p/\mathrm{Im}\ C^{p-1} \longrightarrow C^{p+1}$$

is a Z^{p+2}-isomorphism, its kernel, $H^p(C^{\cdot})$ and its cokernel, $C^{p+1}/\mathrm{Im}\ C^p$, have supports in Z^{p+2}. Furthermore, C^{p+1} lies on the Z^{p+1}/Z^{p+2}-skeleton, so we have a complex with the required properties. To show that it is unique up to unique isomorphism, let $C'^{\cdot}$ be another such augmented complex, defined in degrees $\leq p+1$. Leaving off C'^{p+1}, we have a complex satisfying the conditions of the induction hypothesis, therefore there is a unique isomorphism of augmented complexes $C^{\cdot} \longrightarrow C'^{\cdot}$ defined in degrees $\leq p$. Thus there is a unique isomorphism

$$C^p/\mathrm{Im}\ C^{p-1} \longrightarrow C'^p/\mathrm{Im}\ C'^{p-1}.$$

Now again by the Lemma, this extends to a unique isomorphism of $C^{p+1} \longrightarrow C'^{p+1}$.

Definition. The complex of the above proposition is called the Cousin complex of F (with respect to the filtration $Z^{\cdot}$).

Examples. (1) If F has supports in Z^1, then its Cousin complex is the zero complex.

(2) If F is flasque, then its Cousin complex is given by $C^0 = F/\underline{\Gamma}_{Z^1}(F)$, $C^p = 0$ for $p > 0$.

Lemma 2.4. Under the hypotheses of Proposition 2.3, $\underline{H}^i_{Z^p/Z^{p+1}}(F) = 0$ for all $i > p$.

Proof. Using Motif F, we have only to show that for all $x \in Z^p - Z^{p+1}$, $H^i_x(F) = 0$. But this group can be calculated as the ith derived functor of Γ_x on the topological space $X_{(x)}$ consisting of all generizations of x (i.e., points x' which specialize to x). Indeed, it can be calculated by flasque sheaves (Motif C), and the restriction of a flasque sheaf to $X_{(x)}$ is flasque. Now since $x \notin Z^{p+1}$, the space $X_{(x)}$ has combinatorial dimension $\geq p$, and so $H^i_x(F) = 0$ for $i > p$ [G,II,4.15.2], or [LC 1.12].

Proposition 2.5. Let $X = Z^0 \supseteq Z^1 \supseteq \cdots$ and F be as in Proposition 2.3, and assume furthermore that the filtration is separated (i.e., $\bigcap Z^n = \emptyset$). Then the natural map

$\alpha: F \longrightarrow \underline{H}^0_{Z^0/Z^1}(F)$ makes the complex

$$0 \longrightarrow \underline{H}^0_{Z^0/Z^1}(F) \xrightarrow{d_1^{00}} \underline{H}^1_{Z^1/Z^2}(F) \xrightarrow{d_1^{10}} \underline{H}^2_{Z^2/Z^3}(F) \longrightarrow \cdots$$

of E_1^{p0} terms of the spectral sequence of Motif G into a Cousin complex for F.

Proof. We must check first that α gives an augmentation (i.e., $d_1^{00} \circ \alpha = 0$), and then that the properties (a), (b), and (c) of Proposition 2.3 hold. These properties can be checked at each point separately. Thus for an $x \in X$, we can replace X by the space $X_{(x)}$ of generizations of x, which has finite combinatorial dimension (since the filtration $Z^\cdot$ is separated). In other words, we may assume that the filtration $Z^\cdot$ is finite. To avoid boring the reader, we will check only property (b) which is the hardest.

Given $p > 0$, we wish to show that H^p of the complex above, which is nothing but E_2^{p0} of the spectral sequence of Motif G, has supports in Z^{p+2}. We show, by descending induction on r, that E_r^{p0} has supports in Z^{p+r} for each $r > 1$. For r large enough, $E_r^{p0} = E_\infty^{p0} = 0$, since the abutment of the spectral sequence, E^n, is 0 for $n \neq 0$.

Now let $r > 1$ be given. Then $E_r^{p-r,r-1} = 0$, since $E_r^{pq} = 0$ for $q > 0$ by the Lemma. Therefore we have an exact sequence

$$0 \longrightarrow E_{r+1}^{p0} \longrightarrow E_r^{p0} \xrightarrow{d_r^{p0}} E_r^{p+r,-r+1}.$$

By the induction hypothesis, E_{r+1}^{p0} has supports in Z^{p+r+1}; and $E_1^{p+r,-r+1} = \underline{H}^{p+1}_{Z^{p+r}/Z^{p+r+1}}(F)$, has supports in Z^{p+r}. Thus E_r^{p0} has supports in Z^{p+r}, and in particular, E_2^{p0} has supports in Z^{p+2}.

Now we ask when the Cousin complex is a resolution of F.

Proposition 2.6. Under the hypotheses of Proposition 2.5, the following conditions are equivalent:

(i) $\underline{H}^i_{Z^p}(F) = 0$ for all i,p with $i \neq p$.

(ii) Z^p-depth $F \geq 0$ for all p.

(iii) $\underline{H}^i_{Z^p/Z^{p+1}}(F) = 0$ for all i,p with $i \neq p$.

(iv) The Cousin complex of F is a (flasque) resolution of F.

Proof. (i) $\Longrightarrow$ (ii) by definition of depth: (ii) $\Longrightarrow$ (iii). They are zero for $i > p$ by Lemma 2.4. For $i < p$ we use the exact sequence

$$\underline{H}^i_{Z^p}(F) \longrightarrow \underline{H}^i_{Z^p/Z^{p+1}}(F) \longrightarrow \underline{H}^{i+1}_{Z^{p+1}}(F) .$$

(iii) $\Longrightarrow$ (iv). The condition is pointwise, so as before we can assume the filtration $Z^{\cdot}$ is finite. Then the spectral sequence of Motif G degenerates: $E_1^{pq} = 0$ for $q \neq 0$. That means that the complex E_1^{p0} (which by the previous proposition is the Cousin complex) is a resolution of F.

(iv) $\Longrightarrow$ (i). Let $C^{\cdot}$ be the Cousin complex. Since it is a flasque resolution of F, we may use it to calculate cohomology: $\underline{H}^i_{Z^p} = H^i(\underline{\Gamma}_{Z^p}(C^{\cdot}))$. Since C^p lies on the Z^p/Z^{p+1}-skeleton for all p, $\underline{\Gamma}_{Z^p}(C^{\cdot})$ is the truncated complex $(C^i)_{i \geq p}$. Clearly it has cohomology only in degree p, since the original $C^{\cdot}$ was exact.

Definition. If the equivalent conditions of the proposition are satisfied, we say that F is Cohen-Macaulay (with respect to the filtration $Z^{\cdot}$).

Remark. If X is a locally noetherian prescheme, and Z^p the set of points of codimension $\geq p$ (i.e., points $x \in X$ with $\dim \mathcal{O}_{x,X} \geq p$), and F a coherent sheaf on X with support X, then this notion coincides with the usual definition. Indeed, using condition (iii) and Motif F, F is Cohen-Macaulay if and only if for each $x \in X$, $\operatorname{depth}_{\mathcal{O}_x} F_x \geq \dim \mathcal{O}_x$, which is the usual definition.

Example. Let X be a non-singular locally noetherian scheme, let $Z^{\cdot}$ be the filtration by codimension as above, and let $F = \mathcal{O}_X$. Then F is Cohen-Macaulay (usual sense) [ZS vol. II, App. 6], so the Cousin complex gives a flasque resolution of $\mathcal{O}_X$. Furthermore, the pth component of this complex is isomorphic to $\underline{H}^p_{Z^p/Z^{p+1}}(\mathcal{O}_X)$ which by Motif F is isomorphic to

$$\coprod_{\operatorname{codim} x = p} i_x(H^p_x(\mathcal{O}_X)) .$$

Now for x of codimension p, we know [LC.4.13] that $H^p_x(\mathcal{O}_X)$ is an injective hull over the local ring $\mathcal{O}_x$ of its residue field $k(x)$. Thus our Cousin complex is in fact an injective resolution of $\mathcal{O}_X$, and its component in degree p is isomorphic to a direct sum of sheaves $J(x)$ (see definition in [II §7]) where x is a point of codimension p.

This is an example of the notion of residual complex, which will be studied in more detail in Chapter VI.

§3. Generalization to Complexes.

In this section we generalize the results of the previous section to complexes. In particular, we will discuss Cohen-Macaulay complexes, Gorenstein complexes, and their relations to Cousin complexes.

Throughout this section, X will be a locally noetherian topological space in which every irreducible closed subset has a unique generic point. We will denote by $D^+(X)$ either $D^+(Ab(X))$, the derived category of the category of abelian sheaves on X, or $D^+(Mod(X))$ if X is a locally noetherian prescheme. The results are valid in both cases.

We will consider filtrations $Z^\cdot = (Z^p)_{p\in\mathbb{Z}}$ on X, and will always suppose the following conditions satisfied:

(1) Each Z^p is a subset of X, stable under specialization, and

$$\ldots Z^{p-1} \supseteq Z^p \supseteq Z^{p+1} \supseteq \ldots .$$

(2) Each $x \in Z^p - Z^{p+1}$ is maximal in Z^p (i.e., x is not a proper specialization of any other $x' \in Z^p$).

(3) The filtration is strictly exhaustive, i.e., $X = Z^p$ for some $p \in \mathbb{Z}$.

(4) The filtration is separated, i.e., $\bigcap Z^p = \emptyset$.

<u>Definition</u>. A <u>Cousin complex</u> on X, with respect to the filtration $Z^{\cdot}$, is a complex of sheaves $G^{\cdot}$, such that for each p, G^p lies on the Z^p/Z^{p+1}-skeleton of X. (Cf. definition following Proposition 2.1.) Note that a Cousin complex is necessarily flasque, and bounded below. We denote by <u>$Coz(Z^{\cdot};X)$</u> the category of Cousin complexes and morphisms of complexes. It is an additive category.

<u>Example</u>. The Cousin complex of a sheaf F (as in Proposition 2.3 above) is a Cousin complex.

<u>Definition</u>. Let $F^{\cdot} \in D^+(X)$. Then we denote by $E(F^{\cdot})$ the complex

$$\cdots \longrightarrow \underline{H}^p_{Z^p/Z^{p+1}}(F^{\cdot}) \xrightarrow{d_1^{p0}} \underline{H}^{p+1}_{Z^{p+1}/Z^{p+2}}(F^{\cdot}) \longrightarrow \cdots$$

of E_1^{p0} terms of the spectral sequence of Motif G. We observe by Motif F that the pth term $E^p(F^{\cdot})$ lies on the Z^p/Z^{p+1}-skeleton of X, and so $E(F^{\cdot})$ is a Cousin complex. (Note that even if $F^{\cdot} \in D^b(X)$, the complex $E(F^{\cdot})$ need not be bounded above.)

<u>Proposition 3.1</u>. Let $F^{\cdot} \in D^b(X)$. Then the following conditions are equivalent:

(i) a) $\underline{H}^i_{Z^p}(F^\cdot) = 0$ for $i < p$, and

b) the map

$$\underline{H}^i_{Z^p}(F^\cdot) \longrightarrow H^i(F^\cdot)$$

is surjective for $i = p$ and bijective for $i > p$.

(ii) $\underline{H}^i_{Z^p/Z^{p+1}}(F^\cdot) = 0$ for $i \neq p$.

(iii) There is an isomorphism $\varphi\colon F^\cdot \longrightarrow QE(F^\cdot)$ in $D^+(X)$, where Q is the functor sending a complex to its image in $D^+(X)$ [I §3].

Furthermore, the isomorphism in (iii) can be chosen so that the isomorphisms $H^i(\varphi)\colon H^i(F^\cdot) \longrightarrow H^i(E(F^\cdot))$ are inverse to the isomorphisms $\epsilon_i\colon H^i(E(F^\cdot)) \longrightarrow H^i(F^\cdot)$ determined by the degenerate spectral sequence of Motif G.

Remark. One should beware, however, that the isomorphism φ is not in general unique, and so is not functorial.

Proof of Proposition. (i) $\Longrightarrow$ (ii). This follows immediately from the exact sequences

$$\underline{H}^i_{Z^{p+1}}(F^\cdot) \longrightarrow \underline{H}^i_{Z^p}(F^\cdot) \longrightarrow \underline{H}^i_{Z^p/Z^{p+1}}(F^\cdot) \longrightarrow \underline{H}^{i+1}_{Z^{p+1}}(F^\cdot) \longrightarrow \underline{H}^{i+1}_{Z^p}(F^\cdot)$$

of Motif B.

(ii) $\Longrightarrow$ (i). Condition (i) can be checked pointwise, so as in §2 above, we may assume that the filtration $Z^{\cdot}$ is finite.

a) For $i < p$, the exact sequence of Motif B shows us that

$$\underline{H}^{i}_{Z^{p+1}}(F^{\cdot}) \longrightarrow \underline{H}^{i}_{Z^{p}}(F^{\cdot})$$

is bijective. Therefore by iteration,

$$\underline{H}^{i}_{Z^{p+r}}(F^{\cdot}) \longrightarrow \underline{H}^{i}_{Z^{p}}(F^{\cdot})$$

is bijective for any $r \geq 0$. But for r large enough, $Z^{p+r} = \emptyset$, so $\underline{H}^{i}_{Z^{p}}(F^{\cdot}) = 0$.

Part b) is proved similarly.

(i) + (ii) $\Longrightarrow$ (iii). Since $F^{\cdot} \in D^{b}(X)$, we can find an i_{o} such that $H^{i}(F^{\cdot}) = 0$ for $i \geq i_{o}$. By condition (ii), the spectral sequence of Motif G degenerates, so checking pointwise, we see that $H^{i}(E(F^{\cdot})) = 0$ for $i \geq i_{o}$ also.

We will construct, by descending induction on p, an isomorphism in $D^{+}(X)$ of $\underline{\underline{R\Gamma}}_{Z^{p}}(F^{\cdot})$ with the truncation $Q(\tau_{\geq p} E(F^{\cdot}))$ of $QE(F^{\cdot})$ in degrees $\geq p$ (see [I §7] for notation). Then, since the filtration is strictly exhaustive, we have

$F^{\cdot} = \underline{\underline{R}}\Gamma_{Z^p}(F^{\cdot})$ and $QE(F^{\cdot}) = Q(\tau_{\geq p}E(F^{\cdot}))$ for p small enough, which will give us the isomorphism φ.

For p large enough (say $p \geq i_o$), $\underline{\underline{R}}\Gamma_{Z^p}(F^{\cdot})$ has a unique non-zero cohomology group. Indeed,

$$\underline{H}^i_{Z^p}(F^{\cdot}) = 0 \qquad \text{for } i < p \qquad \text{by (i) a)}$$

and

$$\underline{H}^i_{Z^p}(F^{\cdot}) = H^i(F^{\cdot}) = 0 \quad \text{for } i > p \quad \text{by (i) b),}$$

since $p \geq i_o$. Furthermore, there is an exact sequence

$$(1) \quad 0 \longrightarrow \underline{H}^p_{Z^p}(F^{\cdot}) \longrightarrow \underline{H}^p_{Z^p/Z^{p+1}}(F^{\cdot}) \longrightarrow \underline{H}^{p+1}_{Z^{p+1}}(F^{\cdot}) \longrightarrow 0$$

from Motif B.

At the same time, for $p \geq i_o$, $\tau_{\geq p}E(F^{\cdot})$ has a unique non-zero cohomology group. Indeed, $H^i = 0$ for $i < p$ since the complex is zero in those degrees. $H^i = 0$ for $i > p$, since $p \geq i_o$ (see above). Finally since this is a truncated complex, and $H^p(E(F^{\cdot})) = 0$, there is an exact sequence

$$(2) \qquad 0 \longrightarrow H^p(\tau_{\geq p}E(F^{\cdot})) \longrightarrow E^p(F^{\cdot}) \longrightarrow E^{p+1}(F^{\cdot}) .$$

Now comparing (1) and (2), and using the definition of the complex $E(F^{\cdot})$, and noting that

$$\underline{H}^{p+1}_{Z^{p+1}}(F^{\cdot}) \longrightarrow E^{p+1}(F^{\cdot})$$

is injective, we see that

$$\underline{H}^{p}_{Z^{p}}(F^{\cdot}) \cong H^{p}(\tau_{\geq p}E(F^{\cdot})).$$

Hence there is an isomorphism [I.4.3]

$$\varphi_p\colon\ \underline{\underline{R\Gamma}}_{Z^p}(F^{\cdot}) \xrightarrow{\ \sim\ } Q(\tau_{\geq p}E(F^{\cdot}))$$

in $D^{+}(X)$.

We continue the induction as follows. In terms of the derived category, the spectral sequence of Motif G is expressed as a diagram of triangles

$$\begin{array}{ccccccccc}
\cdots & & \underline{\underline{R\Gamma}}_{Z^p/Z^{p+1}}(F^{\cdot}) & & & & \underline{\underline{R\Gamma}}_{Z^{p-1}/Z^{p}}(F^{\cdot}) & & \cdots \\
 & \swarrow & & \nwarrow & & \swarrow & & \nwarrow & \\
\cdots \longrightarrow & \underline{\underline{R\Gamma}}_{Z^{p+1}}(F^{\cdot}) & \longrightarrow & & \underline{\underline{R\Gamma}}_{Z^{p}}(F^{\cdot}) & \longrightarrow & & \underline{\underline{R\Gamma}}_{Z^{p-1}}(F^{\cdot}) & \longrightarrow \cdots .
\end{array}$$

So suppose by induction we have an isomorphism φ_p of $\underline{\underline{R\Gamma}}_{Z^p}(F^{\cdot})$ with $Q(\tau_{\geq p}E(F^{\cdot}))$, for any p. By (ii), $\underline{\underline{R\Gamma}}_{Z^{p-1}/Z^{p}}(F^{\cdot})$ has a single non-zero cohomology group, so there is an isomorphism

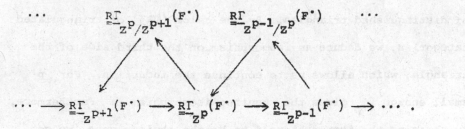

ψ of it in the derived category with the complex consisting of that single sheaf, in the right place [I.4.3], namely $Q(E(F^{\cdot})^{p-1}[-p+1])$. Thus we have a commutative diagram

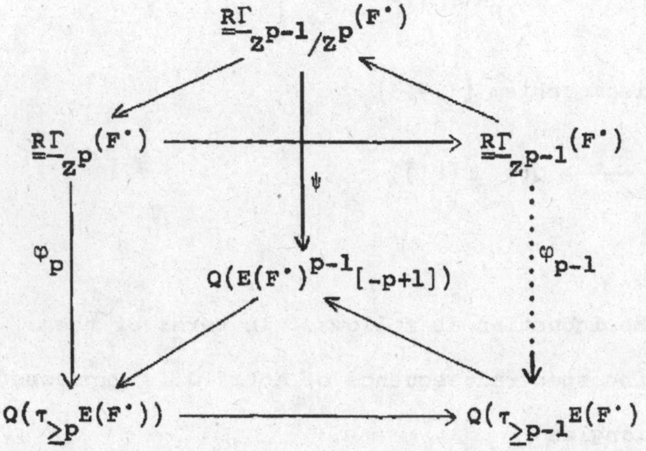

of distinguished triangles. By the axiom (TR3) of triangulated categories, we deduce an isomorphism on the third side of the triangle, which allows us to continue the induction. For p small enough, φ_p gives the required isomorphism φ. Furthermore, by choosing the isomorphisms ψ to be the obvious ones, we get the further condition on the $H^i(\varphi)$.

(iii) $\Longrightarrow$ (ii). We can use the (flasque) complex $E(F^{\cdot})$ to calculate cohomology, by Motif C. Then $\underline{\Gamma}_{Z^p/Z^{p+1}}(E(F^{\cdot})) = E^p(F^{\cdot})$, whence the result.

<u>Definition</u>. If $F^\cdot \in D^b(X)$ satisfies the equivalent conditions of the Proposition, we say that $F^\cdot$ is <u>Cohen-Macaulay</u> with respect to the filtration $Z^\cdot$. If F is a single sheaf, this is the same as the notion of the previous section (Proposition 2.6).

<u>Lemma 3.2</u>. Let $F^\cdot, G^\cdot \in \mathrm{Coz}(Z^\cdot;X)$. Then

a) If f,g are two morphisms of $F^\cdot$ into $G^\cdot$ such that

$$H^i(f),\ H^i(g):\quad H^i(F^\cdot) \longrightarrow H^i(G^\cdot)$$

are the same map for each i, then $f = g$.

b) If $\bar{f}$: $Q(F^\cdot) \longrightarrow Q(G^\cdot)$ is a morphism in $D^+(X)$, and if $G^\cdot$ is a complex of injective sheaves, then $\bar{f}$ is represented by an actual morphism of complexes f: $F^\cdot \longrightarrow G^\cdot$.

<u>Proof</u>. a) By considering $f-g$, we reduce to showing that if f: $F^\cdot \longrightarrow G^\cdot$ induces the zero map on cohomology, then f is the zero map itself. We may assume by translation that $X = Z^o$. Then since $H^o(F^\cdot) \subseteq F^o$, and $H^o(f)$ is the zero map, the map

$$f^o:\ F^o \longrightarrow G^o$$

passes to the quotient to give a map

$$\bar{f}^o \quad B^1(F^\cdot) \longrightarrow G^o.$$

But $B^1(F^\cdot)$ has supports in Z^1, and G^o has Z^1-depth ≥ 1, i.e., $\underline{\Gamma}_{Z^1}(G^o) = 0$, so there are no non-zero maps of $B^1(F^\cdot)$ into G^o. We conclude that $\overline{f}^o$, and hence also f^o, is the zero map.

Thus f^1 maps $Z^1(F^\cdot)$ into 0, hence gives a map

$$\overline{f}^1 \colon B^2(F^\cdot) \longrightarrow G^1 .$$

Proceeding as above we shows that f^p is the zero map for all p.

b) This follows directly from [I.4.5] or [I.4.7], and is included only as a reminder.

Corollary 3.3. Let $F^\cdot, G^\cdot \in \mathrm{Coz}(Z^\cdot;X)$. If $f\colon F^\cdot \longrightarrow G^\cdot$ is a morphism of complexes which induces an isomorphism on the cohomology sheaves, and if $F^\cdot$ is an injective complex, then f is an isomorphism of complexes.

Definition. A complex $F^\cdot \in D^b(X)$ is Gorenstein with respect to the filtration $Z^\cdot$ if it is Cohen-Macaulay, and if $H^i_x(F^\cdot)$ is either zero or injective for each $i \in \mathbb{Z}$, $x \in X$. We denote the (additive) subcategory of $D^b(X)$ of Gorenstein complexes by $D^b(X)_{Gor(Z^\cdot)}$.

Remarks. 1. If X is the spectrum of a local noetherian ring A, and if $Z^{\cdot}$ is the filtration by codimension, and if $F^{\cdot} = A$, then this is the usual notion of a Gorenstein ring [LC 4.14 and Exercise 2 ff].

2. If $F^{\cdot}$ is Gorenstein, then $E(F^{\cdot})$ is an injective Cousin complex, by Motif F.

*Example. We will see in Chapter V that a dualizing complex on a prescheme in Gorenstein with respect to the corresponding filtration by codimension. *

Proposition 3.4. The functor

$$E: \quad D^b(X)_{Gor(Z^{\cdot})} \longrightarrow Icz(Z^{\cdot},X)$$

is an equivalence of categories of the category of Gorenstein complexes with the (additive) category $Icz(Z^{\cdot},X)$ of injective Cousin complexes. Furthermore, the natural functor Q is an inverse to E.

Proof. We must construct isomorphisms of functors

$$\varphi: \quad 1 \longrightarrow QE$$

and

$$\psi: \quad 1 \longrightarrow EQ$$

on the two categories. To construct φ, we choose, for each Gorenstein complex $F^{\cdot} \in D^b(X)$, an isomorphism

$$\varphi(F^{\cdot}):\ F^{\cdot} \longrightarrow QE(F^{\cdot})$$

as in Proposition 3.1 (iii) such that

$$H^i(\varphi(F^{\cdot})):\ H^i(F^{\cdot}) \longrightarrow H^i(E(F^{\cdot}))$$

is inverse to the map derived from the spectral sequence. I claim the collection of morphisms $\{\varphi(F^{\cdot})\}$ is an isomorphism of functors

$$\varphi:\ 1 \longrightarrow QE$$

as required. Indeed, let $f: F^{\cdot} \longrightarrow G^{\cdot}$ be a morphism of Gorenstein complexes in $D^b(X)$. Then we have a diagram

$$\begin{array}{ccc} F^{\cdot} & \xrightarrow{\quad f \quad} & G^{\cdot} \\ {\scriptstyle \varphi(F^{\cdot})}\downarrow & & \downarrow{\scriptstyle \varphi(G^{\cdot})} \\ QE(F^{\cdot}) & \xrightarrow{\quad QE(f) \quad} & QE(G^{\cdot}) \end{array}$$

where $\varphi(F^{\cdot}), \varphi(G^{\cdot})$ are isomorphisms in the derived category. To show this diagram is commutative, we must show that $QE(f) = \varphi(G^{\cdot})f\varphi(F^{\cdot})^{-1}$ in $D^+(X)$. Now by the lemma, part b), $\varphi(G^{\cdot})f\varphi(F^{\cdot})^{-1}$ is represented by an actual morphism of complexes

$$g:\ E(F^{\cdot}) \longrightarrow E(G^{\cdot})\ .$$

Since E is a functor, our condition on $H^i(\varphi(F^\cdot))$ and $H^i(\varphi(G^\cdot))$ shows that $E(f)$ and g have the same effect on cohomology. Hence by the lemma, part a), they are equal. Thus φ is functorial, as required.

For ψ, one need only note that if $F^\cdot \in Coz(Z^\cdot, X)$, then $EQ(F^\cdot) = F^\cdot$ in an obvious way.

Remark. This Proposition will be used in an essential way in Chapter VI in the construction of residual complexes from dualizing complexes.

CHAPTER V. DUALIZING COMPLEXES AND LOCAL DUALITY

§0. Introduction.

In this chapter we discuss dualizing complexes on a locally noetherian prescheme X. A dualizing complex is a complex $R^{\cdot} \in D^{+}(X)$ such that the functor

$$\underline{D}:\ M^{\cdot} \longrightarrow \underline{\underline{R}}\ \underline{Hom}^{\cdot}(M^{\cdot},R^{\cdot})$$

induces an auto-duality of the category $D_c^b(X)$ consisting of those bounded complexes in $D^{+}(X)$ which have coherent cohomology. We will show that a large class of preschemes admits dualizing complexes, and that they are almost uniquely determined.

The notion of dualizing complex will allow us to write the duality theorem in a new way. For example, let X be projective n-space over a field k. Then $R^{\cdot} = \omega[n]$ (where $\omega = \omega_{X/k}$ is the sheaf of relative n-differentials [III §1]) is a dualizing complex for X, and k is a dualizing complex for k. We have a canonical isomorphism $\underline{\underline{R}}f_*(\omega[n]) \cong k$ [III 3.4], and hence, for any complex $F^{\cdot} \in D_c^b(X)$, a homomorphism

$$\theta:\ \underline{\underline{R}}f_*\ \underline{D}_X(F^{\cdot}) \longrightarrow \underline{D}_k(\underline{\underline{R}}f_*(F^{\cdot}))\ ,$$

where $\underline{D}_X$ and $\underline{D}_k$ are the dualizing functors on X and on k, respectively. The duality theorem [III 5.1] says that θ is an isomorphism, i.e., that the dualizing functors on X and on k commute with $\underline{\underline{R}}f_*$.

In the latter part of the chapter we discuss the local duality theorem, as in [LC §2,4,6] or [SGA 62, exposé 2,4]. The proof of the duality theorem given here is different, however.

The reader will see that the existence of a dualizing complex on a locally noetherian prescheme X implies that X has finite Krull dimension. Hence we will give a more general notion, that of a pointwise dualizing complex, to cover the case of infinite Krull dimension.

For convenience we defer the question of existence of dualizing complexes until the end of the chapter (section 10).

§1. Example: Duality for abelian groups.

Let $X = \operatorname{Spec} \mathbb{Z}$. Then the category of quasi-coherent sheaves on X is isomorphic to the category (Ab) of abelian groups. We know that $\mathbb{Q}/\mathbb{Z}$ gives a good duality for finite abelian groups, i.e., that the functor $\operatorname{Hom}(\cdot, \mathbb{Q}/\mathbb{Z})$ is an exact contravariant functor, which, when applied twice to a finite abelian group, gives that group back again. Similarly, the group $\mathbb{Q}$ gives a good notion of duality for finitely generated free abelian groups. Combining the two, we consider the complex

$$\mathbb{Q} \longrightarrow \mathbb{Q}/\mathbb{Z}$$

and work in the derived category $D_c^b(\mathrm{Ab})$, consisting of those bounded complex in $D^+(\mathrm{Ab})$ which have finitely generated cohomology. (c is for "coherent", which means "finitely generated" in our case.) Since the complex above is an injective resolution of $\mathbb{Z}$, hence isomorphic to $\mathbb{Z}$ in the derived category, we arrive at the following proposition.

Proposition 1.1. The functor

$$D:\ M^{\bullet} \longrightarrow \underline{\underline{R}}\operatorname{Hom}^{\bullet}(M^{\bullet},\mathbb{Z})$$

is a contravariant ∂-functor from $D_c^b(Ab)$ into itself, and there is a natural functorial isomorphism

$$\eta:\ id_{D_c^b(Ab)} \xrightarrow{\sim} DD .$$

<u>Proof</u>. Since the cohomology of $D(M^\cdot)$ is $Ext^i(M^\cdot,\mathbb{Z})$, if $M^\cdot \in D_c^b(Ab)$, then so is $D(M^\cdot)$. The natural functorial homomorphism η is defined in Lemma 1.2 below. To show that it is an isomorphism, by taking a free resolution of $M^\cdot$ and applying the lemma on way-out functors [I.7.1], we reduce to the case $M^\cdot = \mathbb{Z}^r$ for some r. Since the functors in question are additive, one reduces to the case $M^\cdot = \mathbb{Z}$, and to complete the proof, one need only observe that $Ext^i(\mathbb{Z},\mathbb{Z}) = \mathbb{Z}$ if $i = 0$, and 0 otherwise.

<u>Lemma 1.2</u>. Let X be a prescheme, let $R^\cdot \in D^+(X)$ and let $F^\cdot \in D(X)$. Then there is a natural functorial homomorphism

$$\eta:\ F^\cdot \longrightarrow \underline{\underline{R}}\ \underline{Hom}^\cdot(\underline{\underline{R}}\ \underline{Hom}^\cdot(F^\cdot,R^\cdot),\ R^\cdot) .$$

<u>Proof</u>. Replacing $R^\cdot$ by an injective resolution, we can erase the $\underline{\underline{R}}$. As usual, we need only define η on <u>objects</u> of D(X), i.e., complexes. Given an index p, and a section $s \in F^p(U)$

over an open set U in X, we must define, for each q, a homomorphism of sheaves

$$\underline{\mathrm{Hom}}^q(F^\cdot,R^\cdot)_U \longrightarrow R_U^{p+q} ,$$

where the subscript U denotes restriction of sheaves to U. Given a homomorphism $f \in \underline{\mathrm{Hom}}^q(F^\cdot,R^\cdot)_U(V)$ defined on an open set $V \subseteq U$, we send it to the section $f(s_V) \in R^{p+q}(V)$, where s_V is the section s, restricted to V. One checks that this indeed does define a morphism of complexes, and that it is functorial in $F^\cdot$.

§2. Dualizing Complexes.

Throughout this section, X will denote a locally noetherian prescheme. We will consider complexes $R^{\cdot} \in D^{+}(X)$ which have coherent cohomology and finite injective dimension, i.e., $R^{\cdot} \in D_c^{+}(X)_{fid}$ (see [I.7.6] and [II.7.20]). Note that if X is quasi-compact, then the condition of finite injective dimension on a complex $R^{\cdot} \in D_c^{+}(X)$ can be checked locally (eg., by [II 7.20(iii)]). Note also that such a complex $R^{\cdot}$ is quasi-isomorphic to a bounded complex of quasi-coherent injective sheaves on X [II 7.20(i)$_{qc}$], and hence (using [II 3.3]) the functor

$$\underline{D}: F^{\cdot} \longrightarrow \underline{\underline{R}}\, \underline{Hom}^{\cdot}(F^{\cdot},R^{\cdot})$$

sends $D_c^b(X)$ into itself, and interchanges $D_c^{+}(X)$ and $D_c^{-}(X)$.

__Definition.__ Let $R^{\cdot} \in D^{+}(X)$. We say that a complex $F^{\cdot} \in D(X)$ is __reflexive__ (with respect to $R^{\cdot}$) if the natural map

$$\eta: F^{\cdot} \longrightarrow \underline{\underline{R}}\, \underline{Hom}^{\cdot}(\underline{\underline{R}}\, \underline{Hom}^{\cdot}(F^{\cdot},R^{\cdot}), R^{\cdot})$$

of Lemma 1.2 is an isomorphism.

<u>Proposition 2.1.</u> Let $R^{\cdot} \in D_c^+(X)_{fid}$. Then the following conditions are equivalent:

(i) Every $F^{\cdot} \in D_c(X)$ is reflexive (with respect to $R^{\cdot}$).

(ii) Every $F^{\cdot} \in D_c^b(X)$ is reflexive.

(iii) Every coherent sheaf on X is reflexive.

(iv) The structure sheaf $\mathcal{O}_X$ is reflexive.

<u>Proof.</u> Clearly (i) $\Longrightarrow$ (ii) $\Longrightarrow$ (iii) $\Longrightarrow$ (iv). To prove (iv) $\Longrightarrow$ (i) we note that the question is local on X. Hence we may assume that X is affine. Then every coherent sheaf on X is the quotient of a free sheaf $\mathcal{O}_X^r$; the functors in question are additive, so the result follows from the Lemma on Way-Out Functors [I.7.1 (ii) and (iv)], taking into account [II 7.20(<u>ii</u>)] which implies that $\underline{\underline{R}}$ <u>Hom</u>$^{\cdot}(\cdot,R^{\cdot})$ is way-out in both directions.

<u>Definition.</u> Let X be a locally noetherian prescheme. A complex $R^{\cdot} \in D_c^+(X)_{fid}$ satisfying the equivalent conditions of the proposition is called a <u>dualizing complex</u> for X.

<u>Definition.</u> The <u>Krull dimension</u> of a locally noetherian prescheme X is the largest n (or $+\infty$) for which there is a chain $Z_o < Z_1 < \cdots < Z_n$ of closed irreducible subsets Z_i of X. It is also the sup of the dimensions of the local rings of X.

Example 2.2. Let X be a regular prescheme of finite Krull dimension. (A regular prescheme is a prescheme X such that all the local rings $\mathcal{O}_{x,X}$ of points of X are regular local rings.) Then the structure sheaf $\mathcal{O}_X$ is a dualizing complex for X.

Indeed, condition (iv) of the proposition is trivial, because $\underline{\mathrm{Ext}}^i(\mathcal{O}_X,\mathcal{O}_X) = \mathcal{O}_X$ for i = 0, and 0 otherwise. Clearly $\mathcal{O}_X \in D_c^+(X)$. To show that $\mathcal{O}_X$ has finite injective dimension, we use condition $\underline{(iii)}_c$ of [I.7.20]. Let n_o be the Krull dimension of X. Then for every coherent sheaf F on X, $\underline{\mathrm{Ext}}^i(F,\mathcal{O}_X) = 0$ for $i > n_o$, because $\underline{\mathrm{Ext}}^i$ commutes with taking stalks, and the local rings $\mathcal{O}_x$ all have cohomological dimension $\leq n_o$, since they are regular local rings [14 (28.2)].

Corollary 2.3. Let $R^\cdot \in D_c^+(X)_{fid}$. Then $R^\cdot$ is dualizing if and only if for every $x \in X$, $R_x^\cdot$ is dualizing on the local scheme Spec $\mathcal{O}_x$. Furthermore, it is sufficient to take the closed points $x \in X$.

Proof. For $x \in X$, we see that the stalk $R_x^\cdot$ has finite injective dimension in $D_c^+(\mathrm{Spec}\ \mathcal{O}_x)$ by using [II 7.20 $(i)_{qc}$] and [II.7.17 (iii)]. Thus the result follows from condition (iv) of the Proposition.

Proposition 2.4. Let $f: X \longrightarrow Y$ be a finite morphism of locally noetherian preschemes, and let $R^{\cdot}$ be a dualizing complex on Y. Then $f^{\flat}R^{\cdot}$ is a dualizing complex on X (cf. [III §6]).

Proof. We may assume that $R^{\cdot}$ is a bounded complex of quasi-coherent injective sheaves on Y. Then

$$f^{\flat}(R^{\cdot}) = \underline{\mathrm{Hom}}_Y(f_*\mathcal{O}_X, R^{\cdot})^{\sim} ,$$

which is a bounded complex of quasi-coherent injectives on X. Hence $f^{\flat}(R^{\cdot}) \in D_c^b(X)_{fid}$. To see that $f^{\flat}(R^{\cdot})$ is dualizing, we apply condition (iv) of the previous proposition, and consider the map

$$\eta: \mathcal{O}_X \longrightarrow \underline{\underline{R}}\ \underline{\mathrm{Hom}}_X^{\cdot}(\underline{\underline{R}}\ \underline{\mathrm{Hom}}_X^{\cdot}(\mathcal{O}_X, f^{\flat}(R^{\cdot})), f^{\flat}(R^{\cdot})).$$

Applying $\underline{\underline{R}}f_*$ to this map, and using duality for a finite morphism [III 6.7] twice, we see that $\underline{\underline{R}}f_*(\eta)$ is an isomorphism. But we are dealing with quasi-coherent sheaves, and f_* is faithful, so η is an isomorphism.

Remarks. 1. We will see later that if f is a smooth morphism, then $f^{\#}$ of a dualizing complex is dualizing (Theorem 8.3 below).

2. This result, joined with the example above, shows that any closed subscheme of a regular prescheme of finite Krull dimension admits a dualizing complex.

<u>Proposition 2.5</u>. A complex $R^{\cdot} \in D_c^+(X)_{fid}$ is dualizing if and only if for every closed point $x \in X$, the sheaf $k(x)$, consisting of the residue field $k(x)$ at the point x and zero elsewhere, is reflexive.

<u>Proof</u>. 1) Using Corollary 2.3, we reduce to the case where X is the spectrum of a local noetherian ring A, with residue field k. The assumption is then that k is reflexive, and we wish to show for every A-module of finite type M (in particular, for M = A), that M is reflexive on X.

2) Using induction on the length, we show first that every A-module M of finite length is reflexive. Indeed, for length M = 1, we have assumed it. For length M > 1, one can write a short exact sequence

$$0 \longrightarrow M' \longrightarrow M \longrightarrow M'' \longrightarrow 0$$

where M' and M" have length < length M. Then by the induction hypothesis M' and M" are reflexive, and by long exact sequences

(DD is a ∂-functor!) and the five-lemma, M is reflexive.

3) Now we show that any module of finite type M is reflexive, using induction on the dimension of M. If dim M = 0, then M has finite length, and M is reflexive by 2) above. Let M be a module of finite type. Let M' be the submodule of elements with support $\mathfrak{m}$, where $\mathfrak{m}$ is the maximal ideal of A. Then M' has finite length, so it is sufficient to consider M/M'. In other words, we have reduced to the case $\mathfrak{m} \notin$ Ass M.

Let $t \in \mathfrak{m}$ be an element which is not a zero-divisor in M, so that we have an exact sequence

$$0 \longrightarrow M \xrightarrow{t} M \longrightarrow M/tM \longrightarrow 0 .$$

Now dim M/tM < dim M, so by the induction hypothesis M/tM is reflexive. Therefore, for each $i \neq 0$, we have an exact sequence

$$H^i(DD(M)) \xrightarrow{t} H^i(DD(M)) \longrightarrow 0.$$

Now these H^i are A-modules of finite type, and $t \in \mathfrak{m}$, so by Nakayama's lemma, $H^i(DD(M)) = 0$ for $i \neq 0$.

Consider the commutative diagram

$$\begin{array}{ccccccccc}
0 & \longrightarrow & M & \xrightarrow{t} & M & \longrightarrow & M/tM & \longrightarrow & 0 \\
 & & \Big| & & \Big| \alpha & & \Big| \approx & & \\
0 & \longrightarrow & H^0(DD(M) & \xrightarrow{t} & H^0(DD(M)) & \longrightarrow & H^0(DD(M/tM)) & \longrightarrow & 0
\end{array}$$

with exact rows. A diagram chase shows that

$$H^{o}(DD(M)) = \alpha(M) + tH^{o}(DD(M)) ,$$

so by Nakayama's Lemma, α is surjective. To show that α is injective, let $x \in M$. Choose n so large that $x \notin t^{n}M$, and draw the same diagram as above, but with t^{n} in place of t. Then x does not become 0 in $M/t^{n}M$, and $M/t^{n}M \cong H^{o}(DD(M/t^{n}M))$, so $\alpha(x) \neq 0$.

Thus M is reflexive. Taking M = A, we find that $R^{\cdot}$ is a dualizing complex. q.e.d.

<u>Proposition 2.6.</u> Let X be a locally noetherian prescheme, let $R^{\cdot} \in D_{c}^{+}(X)_{fid}$ be a dualizing complex, and let D be the functor $\underline{\underline{R}}\ \underline{Hom}^{\cdot}(\cdot, R^{\cdot})$. Then

a). A complex $G^{\cdot} \in D_{c}^{b}(X)$ has finite Tor-dimension [II 4.2] if and only if $D(G^{\cdot})$ has finite injective dimension.

b). There is a functorial isomorphism

$$D(\underline{\underline{R}}\ \underline{Hom}^{\cdot}(F^{\cdot},G^{\cdot})) \xrightarrow{\sim} F^{\cdot} \underline{\underline{\otimes}} D(G^{\cdot})$$

for $F^{\cdot} \in D_{c}^{-}(X)$ and $G^{\cdot} \in D^{+}(X)$, or for $F^{\cdot} \in D_{c}(X)$ and $G^{\cdot} \in D^{+}(X)_{fid}$.

c). There is a functorial isomorphism

$$D(F^{\cdot} \underset{=}{\otimes} G^{\cdot}) \xrightarrow{\sim} \underset{=}{R}\, \underline{Hom}^{\cdot}(F^{\cdot}, D(G^{\cdot}))$$

for $F^{\cdot}$ and $G^{\cdot} \in D_c^-(X)$, or for $F^{\cdot} \in D_c(X)$ and $G^{\cdot} \in D_c^b(X)_{fTd}$.

Proof. The natural map of sheaves

$$F \otimes \underline{Hom}(G,R) \longrightarrow \underline{Hom}\,(\underline{Hom}(F,G),R)$$

gives rise to a morphism of functors on the derived category,

$$(*) \qquad F^{\cdot} \underset{=}{\otimes} \underset{=}{R}\,\underline{Hom}^{\cdot}(G^{\cdot},R^{\cdot}) \longrightarrow \underset{=}{R}\, \underline{Hom}^{\cdot}(\underset{=}{R}\, \underline{Hom}^{\cdot}(F^{\cdot},G^{\cdot}),R^{\cdot}),$$

provided either

1) $F^{\cdot} \in D^-(X)$, $G^{\cdot} \in D^+(X)$ and $R^{\cdot} \in D^+(X)_{fid}$, or

2) $F^{\cdot} \in D(X)$, $G^{\cdot},R^{\cdot} \in D^+(X)$, and $\underset{=}{R}\, \underline{Hom}^{\cdot}(G^{\cdot},R^{\cdot}) \in D^b(X)_{fTd}$.

If under the set of conditions 1), we assume furthermore that $F^{\cdot} \in D_c^-(X)$, then by [I.7.1], the morphism (*) is an isomorphism. Indeed, we reduce to the case $F = \mathcal{O}_X$, which is trivial. Taking the inverse isomorphism gives b) under the first set of hypotheses. As a corollary, we see that if $G^{\cdot} \in D^+(X)$ has finite injective dimension, then $D(G^{\cdot})$ has finite Tor-dimension. Indeed, by the Remark following [II 4.2] it is enough to show that there is an integer n_o such that $\underline{Tor}_i(F,G^{\cdot}) = 0$ for all $i > n_o$

and all coherent sheaves F on X, and this follows from b). Thus we have established one half of a). Now if $G^{\cdot} \in D^{+}(X)_{fid}$, then as just remarked, $D(G^{\cdot}) \in D^{b}(X)_{fTd}$, so by 2) and [I.7.1] again, we obtain the isomorphism b) under the second set of conditions.

For statement c), let $F^{\cdot}, G^{\cdot} \in D_c^-(X)$, apply b) to $F^{\cdot}$ and $D(G^{\cdot})$, and apply D to the resulting isomorphism. Then since $D^2 = \mathrm{id}$ on $D_c(X)$, we have the required isomorphism under the first set of conditions. As a corollary we deduce (using [II 7.20 (iii)$_c$]) that if $G^{\cdot} \in D_c^b(X)$ has finite Tor-dimension, then $D(G^{\cdot})$ has finite injective dimension. This gives the other half of a), and applying D to the second half of b) gives the second half of c). q.e.d.

§3. The Uniqueness of the Dualizing Complex.

Theorem 3.1. Let X be a connected, locally noetherian prescheme, and let $R^{\cdot}$ be a dualizing complex on X. Let $R'^{\cdot} \in D_c^+(X)$ be any other complex. Then $R'^{\cdot}$ is a dualizing complex if and only if there exists an invertible sheaf L on X, and an integer n, such that

$$R'^{\cdot} \cong R^{\cdot} \underline{\underline{\otimes}} L[n] .$$

Furthermore, L and n are determined uniquely to within isomorphism.

Proof. 1) Observe that such an $R'^{\cdot}$ is a dualizing complex. Indeed, we may assume that $R^{\cdot}$ is a bounded complex of quasi-coherent injectives. Then $R^{\cdot} \otimes L[n]$ is also a bounded complex of injectives, since the property of being injective is local [II 7.16], and L is locally free. $R'^{\cdot}$ is dualizing by Corollary 2.3.

2) Observe that L and n are uniquely determined by the identity $L[n] \cong \underline{\underline{R}}\, \underline{\mathrm{Hom}}^{\cdot}(R^{\cdot},R'^{\cdot})$.

3) Now we show that existence of L and n, assuming that $R'^{\cdot}$ is a dualizing complex. Let D and D' be the dualizing

functors corresponding to $R^\cdot$ and $R'^\cdot$. Define

$$L^\cdot = D'D(\mathcal{O}_X) = \underline{\underline{R}}\ \underline{Hom}^\cdot(R^\cdot, R'^\cdot) .$$

We will prove that $L^\cdot$ is isomorphic, in $D^+(X)$, to a complex of the form $L[n]$, where L is an invertible sheaf, and that $R'^\cdot \cong R^\cdot \underline{\underline{\otimes}} L^\cdot$.

Lemma 3.2. There is a natural functorial isomorphism, for all $M^\cdot \in D_c^-(X)$,

$$M^\cdot \underline{\underline{\otimes}} L^\cdot \xrightarrow{\sim} D'D(M^\cdot) .$$

(Note that this tensor product makes sense since $L^\cdot \in D_c^-(X)$, cf. [II §4].)

Proof. As in Lemma 1.2 above, one defines a functorial homomorphism

$$M^\cdot \underline{\underline{\otimes}}\ \underline{\underline{R}}\ \underline{Hom}^\cdot(R^\cdot, R'^\cdot) \longrightarrow \underline{\underline{R}}\ \underline{Hom}^\cdot(\underline{\underline{R}}\ \underline{Hom}^\cdot(M^\cdot, R^\cdot), R'^\cdot).$$

Since it is an isomorphism for $M^\cdot = \mathcal{O}_X$, by the lemma on way-out functors, it is an isomorphism for all $M^\cdot \in D_c^-(X)$.

We now define $L'^\cdot = DD'(\mathcal{O}_X)$, and use the lemma to deduce that

$$L^\cdot \underline{\underline{\otimes}} L'^\cdot = D'D(L'^\cdot) = D'DDD'(\mathcal{O}_X) = \mathcal{O}_X .$$

Lemma 3.3. Let X be a connected locally noetherian prescheme, and let $L^{\cdot}, L'^{\cdot} \in D_c^-(X)$ be such that $L^{\cdot} \underline{\underline{\otimes}} L'^{\cdot} \cong \mathcal{O}_X$. Then $L^{\cdot} = L[n]$ for some invertible sheaf L, and $L'^{\cdot} = L^{\vee}[-n]$.

Proof. Since X is connected, we reduce immediately to the case where X is the spectrum of a local noetherian ring A, and we wish to show that $L^{\cdot} = \mathcal{O}_X[n]$ for some n. Let p be the largest integer such that $H^p(L^{\cdot}) \neq 0$, and let q be the largest integer such that $H^q(L'^{\cdot}) \neq 0$. Then one sees easily that $p+q$ is the largest integer for which $H^n(L^{\cdot} \underline{\underline{\otimes}} L'^{\cdot}) \neq 0$, and $H^{p+q}(L^{\cdot} \underline{\underline{\otimes}} L'^{\cdot}) = H^p(L^{\cdot}) \otimes H^q(L'^{\cdot})$. Therefore $p+q = 0$, and $H^p(L^{\cdot}) \otimes H^q(L'^{\cdot}) = \mathcal{O}_X$. It follows, by a simple lemma on modules of finite type over a local noetherian ring [cf. EGA, Ch. O, 5.4.3], that $H^p(L^{\cdot}) \cong H^q(L'^{\cdot}) \cong \mathcal{O}_X$. Now since $\mathcal{O}_X$ is locally free, one shows that $L^{\cdot} = L_1^{\cdot} \oplus \mathcal{O}_X[-p]$, where $L_1^{\cdot}$ has cohomology only in dimensions $< p$. Similarly, $L'^{\cdot} = L_1'^{\cdot} \oplus \mathcal{O}_X[p]$. Thus from $L^{\cdot} \underline{\underline{\otimes}} L'^{\cdot} \cong \mathcal{O}_X$ we deduce

$$L_1^{\cdot}[p] \oplus L_1'^{\cdot}[-p] \oplus (L_1^{\cdot} \underline{\underline{\otimes}} L_1'^{\cdot}) = 0$$

from which it follows that $L_1^{\cdot}$ and $L_1'^{\cdot}$ are 0 in $D_c^-(X)$, so that $L^{\cdot} \cong \mathcal{O}_X[-p]$, and $L'^{\cdot} \cong \mathcal{O}_X[p]$, as required.

This proves the lemma, so we have only to show that $R'^{\cdot} = R^{\cdot} \underline{\underline{\otimes}} L^{\cdot}$. Indeed, $R^{\cdot} \in D_c^b(X)$, so applying Lemma 3.2,

$$R^{\cdot} \underline{\underline{\otimes}} L^{\cdot} = D'D(R^{\cdot}) = D'(\mathcal{O}_X) = R'^{\cdot} \quad . \qquad \text{q.e.d. Theorem.}$$

We now give an application of the uniqueness theorem.

<u>Proposition 3.4</u>. Let A be a noetherian local ring with residue field k, and let $R^{\cdot} \in D_c^+(A)$. Then $R^{\cdot}$ is dualizing if and only if there is an integer d such that

$$\mathrm{Ext}^i(k,R^{\cdot}) = \begin{cases} 0 & \text{for } i \neq d \\ k & \text{for } i = d. \end{cases}$$

(Here we write $D^+(A)$ for the derived category of the category of A-modules. The subscript c denotes complexes whose cohomology modules are of finite type. We carry over the definitions and results of the previous sections to this case. (Cf. [II.7.19] which ensures that we will not get into trouble.))

<u>Proof</u>. First suppose that $R^{\cdot}$ is dualizing, and let j: Spec k $\longrightarrow$ Spec A be the inclusion. Then by Proposition 2.4,

$$j^{\flat}(R^{\cdot}) = \underline{\underline{R}}\,\mathrm{Hom}^{\cdot}(k,R^{\cdot})$$

is a dualizing complex on k. But k itself is a dualizing complex on k, so by the Uniqueness Theorem above, $j^{\flat}(R^{\cdot}) \cong k[-d]$ for some d. This d will do.

For the converse, it is clear that k is reflexive, so by Proposition 2.5 we have only to show that $R^{\cdot}$ has finite injective dimension. Indeed, we will show by induction on $r = \dim M$, for any A-module M of finite type, that

$$\mathrm{Ext}^i(M,R^{\cdot}) = 0 \quad \text{for} \quad i \notin [d-r,d].$$

Indeed, for dim M = 0, M is of finite length, and our statement follows from the case M = k by induction on the length of M.

For dim M = r > 0, we induct as in the proof of Proposition 2.5. First we may assume that $\mathfrak{m} \notin \mathrm{Ass}\, M$, where $\mathfrak{m}$ is the maximal ideal of A. Then choosing $t \in \mathfrak{m}$ with t a non-zero-divisor in M, we have

$$0 \longrightarrow M \xrightarrow{t} M \longrightarrow M/tM \longrightarrow 0$$

and $\dim M/tM < r$. Then for $i \notin [d-r+1,d]$ we have $\mathrm{Ext}^i(M/tM,R^{\cdot})=0$, hence for $i \notin [d-r,d]$ we have from the long exact sequence of Ext's,

$$\mathrm{Ext}^i(M,R^{\cdot}) \xrightarrow{t} \mathrm{Ext}^i(M,R^{\cdot}) \longrightarrow 0 .$$

It follows by Nakayama's lemma that $\mathrm{Ext}^i(M,R^{\cdot}) = 0$. q.e.d.

<u>Corollary 3.5</u>. Let A be a noetherian local ring, and let $R^{\cdot} \in D_c^{+}(A)$. Then $R^{\cdot}$ is dualizing on A if and only if $R^{\cdot} \otimes_A \hat{A}$ is dualizing on the completion $\hat{A}$ of A.

<u>Proof</u>. Clearly $R^{\cdot} \otimes_A \hat{A} \in D_c^{+}(\hat{A})$. Furthermore,

$$\mathrm{Ext}^i_{\hat{A}}(k, R^{\cdot}\otimes_A\hat{A}) = \mathrm{Ext}^i_A(k,R^{\cdot}) \otimes_A \hat{A}$$

for all i. Thus the "only if" implication is clear. For the "if" implication, note that the functor $\otimes_A\hat{A}$ is faithfully flat, so if M is a non-zero A-module, then $M \otimes_A \hat{A}$ is non-zero. Furthermore, if M is an A-module, such that $M \otimes_A \hat{A} \cong k$, then $M \cong k$. (Indeed, map M to k via the natural map $M \longrightarrow M \otimes_A \hat{A}$. This map becomes an isomorphism upon tensoring with $\hat{A}$, hence was an isomorphism). The result thus follows from the proposition.

§4. Local Cohomology on a Prescheme.

Let X be a prescheme, Y a closed subset, and F a sheaf of $\mathcal{O}_X$-modules. We interpret the local cohomology groups $H^i_Y(F)$ in terms of Ext's.

For each $n \geq 1$, let Y_n be the subscheme of X defined by the sheaf of ideals I^n_Y, where I_Y is the sheaf of ideals of some scheme structure on Y. Then for each n we have a natural map

$$\mathrm{Hom}_{\mathcal{O}_X}(\mathcal{O}_{Y_n}, F) \longrightarrow \Gamma_Y(F) ,$$

since an element of the first group is given by a global section of F which is annihilated by I^n_Y, and hence has support on Y. Taking the direct limit as n varies, we deduce a map

$$\varinjlim_n \mathrm{Hom}(\mathcal{O}_{Y_n}, F) \longrightarrow \Gamma_Y(F).$$

Taking derived functors, we deduce a map of functors from $D^+(X)$ to $D^+(Ab)$

$$(*) \qquad \underline{\underline{R}} \varinjlim_n \mathrm{Hom}(\mathcal{O}_{Y_n}, F^{\cdot}) \longrightarrow \underline{\underline{R}}\, \Gamma_Y(F^{\cdot}) .$$

Theorem 4.1. If X is a noetherian prescheme, and $F^{\cdot} \in D^+_{qc}(X)$, i.e., the complex $F^{\cdot}$ has quasi-coherent cohomology (cf. [II §17]), then $(*)$ is an isomorphism.

Remarks. 1. Since there are enough flasque $\mathcal{O}_X$-modules, $\underline{\underline{R}}\,\Gamma_Y(F^{\cdot})$ is the same as if one had calculated it in the category D^+ (abelian sheaves on X).

2. The derived category does not have direct limits, so we cannot write $\varinjlim_n \underline{\underline{R}}\,\mathrm{Hom}(\mathcal{O}_{Y_n}, F^{\cdot})$, as one is tempted to. However, upon descending from the derived category this difficulty disappears, and we have the Corollary below.

Corollary 4.2. Under the hypotheses of the theorem, the map induced on cohomology,

$$\varinjlim_n \mathrm{Ext}^i(\mathcal{O}_{Y_n}, F^{\cdot}) \longrightarrow H^i_Y(F^{\cdot})$$

is an isomorphism, for all i.

Proofs. If X is noetherian, and F quasi-coherent, then the map

$$\varinjlim_n \mathrm{Hom}(\mathcal{O}_{Y_n}, F) \longrightarrow \Gamma_Y(F)$$

is an isomorphism, since every section of F with support on Y will be annihilated by some power of I_Y. Therefore, by the lemma on Way-Out Functors [I.7.1], the morphism of derived functors (*) is an isomorphism for $F^{\cdot} \in D^+_{qc}(X)$.

For the Corollary, just observe that taking cohomology of complexes commutes with direct limits.

Corollary 4.3. If X is locally noetherian, and $F^{\cdot} \in D^{+}_{qc}(X)$, then the analogous maps

$$\underline{\underline{R}}\ \underset{n}{\varinjlim}\ \underline{\mathrm{Hom}}(\mathcal{O}_{Y_n}, F^{\cdot}) \longrightarrow \underline{\underline{R}}\ \underline{\Gamma}_Y(F^{\cdot})$$

and

$$\underset{n}{\varinjlim}\ \underline{\mathrm{Ext}}^i(\mathcal{O}_{Y_n}, F^{\cdot}) \longrightarrow \underline{H}^i_Y(F^{\cdot})$$

are isomorphisms.

Proof. By definition of the derived category, it is sufficient to prove the second. This follows from Corollary 4.2, since $\underline{\mathrm{Ext}}$ and $\underline{H}_Y$ are sheaves associated to presheaves given by Ext and H_Y.

§5. Dualizing functors on a local noetherian ring.

In this section we recall without proof the definition and some properties of dualizing functors on a local noetherian ring A, with maximal ideal $\mathfrak{m}$. For proofs, see [LC,§4].

Proposition 5.1. Let I be an injective hull of the residue field k of A, and denote by T the functor $\mathrm{Hom}(\cdot, I)$. Then for every A-module M of finite length, the natural map

$$M \longrightarrow T\,T(M)$$

is an isomorphism.

Definition. A contravariant additive functor T from the category $\mathcal{C}_f$ of A-modules of finite length into itself is called a dualizing functor for A at $\mathfrak{m}$ if there exists an injective hull I of k and an isomorphism of functors $T \cong \mathrm{Hom}(\cdot, I)$.

Proposition 5.2. A contravariant additive functor T from $\mathcal{C}_f$ into itself is a dualizing functor if and only if it is exact, and $T(k) \cong k$. In that case one can take $I = \varinjlim_n T(A/\mathfrak{m}^n)$, and then there is a canonical isomorphism $T \cong \mathrm{Hom}(\cdot, I)$.

Proposition 5.3. If I is an injective hull of k, then the functor $\mathrm{Hom}(\cdot, I)$ gives an anti-isomorphism of the category (dcc) of A-modules with descending chain condition (we call them A-modules of co-finite type) and the category (acc) of modules of finite type over the completion $\hat{A}$ of A.

§6. Local Duality.

Throughout this section we let A be a noetherian local ring, with maximal ideal $\mathfrak{m}$, and residue field k, and let $R^{\cdot}$ be a dualizing complex on $X = \operatorname{Spec} A$ (cf. §2). Recall Proposition 3.4 which gives a necessary and sufficient condition for a complex $R^{\cdot} \in D_c^+(A)$ to be dualizing.

Definition. We say that the dualizing complex $R^{\cdot}$ is normalized if the integer d of Proposition 3.4 is zero.

Remark. Since the translate of a dualizing complex is again one, we can normalize by translation.

Proposition 6.1. If $R^{\cdot}$ is normalized, then $\underline{\underline{R}}\Gamma_x(R^{\cdot})$, where x is the closed point of Spec A, is isomorphic in the derived category to an injective hull I of k.

Proof. The cohomology of $\underline{\underline{R}}\,\Gamma_x(R^{\cdot})$ is

$$H_x^i(R^{\cdot}) \cong \varinjlim_n \operatorname{Ext}^i(A/\mathfrak{m}^n, R^{\cdot})$$

by Corollary 4.2 above. On the other hand, we saw from the proof of Proposition 3.4 that $\operatorname{Ext}^i(M,R^{\cdot}) = 0$ for $i \neq 0$ and all M of finite length. Furthermore, by Proposition 5.2, the functor $\operatorname{Ext}^0(\cdot,R^{\cdot})$ is a dualizing functor, and corresponds to the injective hull of k given by

$$I = \varinjlim_n \mathrm{Ext}^0(A/\mathfrak{m}^n, R^\cdot).$$

So we see that $H_x^i(R^\cdot) = 0$ for $i \neq 0$, and $H_x^0(R^\cdot) = I$. Therefore $\underline{\underline{R}}\,\Gamma_x(R^\cdot) \cong I$ in the derived category [I.4.3].

Now let M be an A-module. Then there is a natural homomorphism

$$\Gamma_x(M) \longrightarrow \mathrm{Hom}(\mathrm{Hom}(M,R^\cdot),\ \Gamma_x(R^\cdot)).$$

This gives rise to a morphism of functors on $D^+(A)$,

$$\underline{\underline{R}}\,\Gamma_x(M^\cdot) \longrightarrow \underline{\underline{R}}\,\mathrm{Hom}(\underline{\underline{R}}\,\mathrm{Hom}(M^\cdot,R^\cdot),\ \underline{\underline{R}}\,\Gamma_x(R^\cdot)),$$

since we can take $R^\cdot$ to be an injective complex, in which case $\underline{\underline{R}}\,\Gamma_x(R^\cdot) = \Gamma_x(R^\cdot)$ is also injective. Indeed, whenever J is an injective A-module, and Y a closed subset of Spec A, then $\Gamma_Y(J)$ is injective (cf. [II §7] for the structure of injective A-modules). We now assume (without loss of generality) that $R^\cdot$ is normalized, and so, by the Proposition, have a morphism of functors

$$\theta\colon\ \underline{\underline{R}}\,\Gamma_x(M^\cdot) \longrightarrow \mathrm{Hom}(\underline{\underline{R}}\,\mathrm{Hom}(M^\cdot,R^\cdot),\ I).$$

(Following the usual conventions, we do not write $\underline{\underline{R}}$ before a functor which is already exact.)

Theorem 6.2 (Local Duality). Let A be a local noetherian ring, let $R^{\cdot}$ be a normalized dualizing complex, let I be the corresponding injective hull of k, and let $M^{\cdot} \in D_c^{+}(A)$ (i.e., the cohomology modules of $M^{\cdot}$ are of finite type). Then θ as defined above is an isomorphism.

Corollary 6.3. With A, $R^{\cdot}$, I as above, let M be a module of finite type. Then the homomorphisms

$$\theta^i: \quad H_x^i(M) \longrightarrow \mathrm{Hom}\,(\mathrm{Ext}^{-i}(M,R^{\cdot}), I)$$

induced by θ are isomorphisms.

Proof. Using the Lemma on Way-out Functors [I.7.1] one sees that the Corollary is equivalent to the theorem. To prove the Corollary, note that θ^i is an isomorphism for $M = k$, because $H_x^i(k) = 0$ for $i \neq 0$, and $H_x^0(k) = k$; note also that for all M of finite type, both sides are modules with support at x. Thus we are reduced to proving the following

LEMMA 6.4. Let $\theta^i: S^i \longrightarrow T^i$ be a morphism of covariant cohomological functors on the category of modules over a noetherian ring A. Assume

(i) for every maximal ideal $\mathfrak{m} \subseteq A$, $\theta^i(A/\mathfrak{m})$ is an isomorphism, and

(ii) for every non-maximal prime ideal $\mathfrak{p} \subseteq A$, $S^i(A/\mathfrak{p})$ and $T^i(A/\mathfrak{p})$ have support $<$ Supp $A/\mathfrak{p}$.

Then $\theta^i(M)$ is an isomorphism for every A-module M of finite type.

Proof. By assumption, $\theta^i(A/\mathfrak{m})$ is an isomorphism for every maximal ideal $\mathfrak{m}$. If M is any module of finite type, then M admits a finite filtration each of whose quotients is of the form $A/\mathfrak{p}$ with $\mathfrak{p}$ prime. Thus by the 5-lemma we reduce to the case $M = A/\mathfrak{p}$ with $\mathfrak{p}$ not maximal. By noetherian induction, we may assume that $\theta^i(M')$ is an isomorphism for every M' of finite type with support $<$ Supp $(A/\mathfrak{p})$.

Now for each $f \in M = A/\mathfrak{p}$, $f \neq 0$, we consider the exact sequence

$$0 \longrightarrow M \xrightarrow{f} M \longrightarrow M/fM \longrightarrow 0 ,$$

and apply the functors S^i, T^i to it, splitting it as follows:

$$\begin{array}{ccccccccc} 0 & \longrightarrow & K_f & \longrightarrow & S^i(M) & \xrightarrow{f} & S^i(M) & \longrightarrow & \cdots \\ & & \downarrow{\scriptstyle\alpha} & & \downarrow{\scriptstyle\beta} & & \downarrow & & \\ 0 & \longrightarrow & L_f & \longrightarrow & T^i(M) & \xrightarrow{f} & T^i(M) & \longrightarrow & \cdots \end{array}$$

and

$$\begin{array}{ccccccc}\cdots \longrightarrow S^{i-1}(M) & \longrightarrow & S^{i-1}(M/fM) & \longrightarrow & K_f & \longrightarrow & 0 \\ \downarrow \gamma & & \downarrow \delta & & \downarrow \alpha & & \\ \cdots \longrightarrow T^{i-1}(M) & \longrightarrow & T^{i-1}(M/fM) & \longrightarrow & L_f & \longrightarrow & 0 .\end{array}$$

Now by our induction hypothesis, δ is an isomorphism, so α is surjective. Since $\operatorname{Supp} S^i(M) < \operatorname{Supp} M$ by our assumption (ii), each element of $S^i(M)$ is annihilated by some non-zero $f \in M$. Thus $S^i(M) = \bigcup K_f$ as f ranges over the non-zero elements of M. Similarly $T^i(M) = \bigcup L_f$. Therefore β is surjective. This is true for each i, so also γ is surjective. But that implies α is injective, so α is an isomorphism. Then, as above, β is also an isomorphism. q.e.d.

Corollary 6.5. For M of finite type, the modules $H^i_x(M)$ are of co-finite type (see Proposition 5.3), and we have a functorial isomorphism

$$\operatorname{Ext}^i(M,R^{\cdot})^{\wedge} \longrightarrow \operatorname{Hom}(H_x^{-i}(M), I)$$

of $\hat{A}$-modules.

Proof. Apply the functor Hom $(\cdot,I)$ to the isomorphism θ^i, and use the fact that for any A-module of finite type M, this functor applied twice gives the completion $\hat{M}$. [LC p. 61].

Remark. If A is a regular local ring, then A itself is a dualizing complex for A (see Example 2.2), and we recover the old local duality theorem [LC 6.3]. This is also true more generally for Gorenstein local rings (Theorem 9.1 below).

§7. Application to Dualizing Complexes.

Let X be a locally noetherian prescheme, and let $R^{\cdot}$ be a dualizing complex on X. Then for each point $x \in X$, $R_x^{\cdot}$ is a dualizing complex for the local ring $\mathcal{O}_x$ (Corollary 2.3). Thus by Proposition 3.4 we can find an integer $d(x)$ such that

$$\mathrm{Ext}^i_{\mathcal{O}_x}(k(x), R_x^{\cdot}) = \begin{cases} 0 & \text{for } i \neq d(x) \\ k(x) & \text{for } i = d(x) . \end{cases}$$

Proposition 7.1. With the hypotheses above, if $x \longrightarrow y$ is an immediate specialization, then

$$d(y) = d(x) + 1.$$

Proof. Since the question is local around y, we may assume that $X = \mathrm{Spec}\ \mathcal{O}_y$.

Let Z be the reduced induced subscheme structure on $\overline{\{x\}}$, and let $j: Z \longrightarrow X$ be the inclusion. Then $j^{\flat}R^{\cdot}$ is dualizing on Z by Proposition 2.4, and for any $\mathcal{O}_Z$-module F we have

$$\mathrm{Ext}^i_{\mathcal{O}_Z}(F, j^{\flat}R^{\cdot}) = \mathrm{Ext}^i_{\mathcal{O}_X}(j_*F, R^{\cdot})$$

by duality for j [III 6.7]. Thus $d(x)$, $d(y)$ are the same if calculated on Z with $j^{\flat}R^{\cdot}$, and so we reduce to the case $X = Z$ integral local of dimension 1, with generic point x and closed point y.

By translating, we may assume that $R^{\cdot}$ is normalized, i.e., $d(y) = 0$. To calculate $d(x)$ we consider $\mathrm{Ext}^i_{k(x)}(k(x),R^{\cdot}) = \mathrm{Ext}^i_A(A,R^{\cdot}) \otimes_A k(x)$, where $X = \mathrm{Spec}\, A$. Now by the local duality theorem, $\mathrm{Ext}^i_A(A,R^{\cdot})$ is dual to $H^{-i}_y(A)$. Since A is a domain, $H^0_y(A) = 0$. For $p \neq 0,1$, $H^p_y(A) = 0$ since A is of dimension 1 [LC 1.12]. Thus $\mathrm{Ext}^i(A,R^{\cdot}) = 0$ for $i \neq -1$, and so $d(x) = -1$, as required.

Definition. An integer-valued function d on a prescheme X with the property of the proposition will be called a codimension function on X.

Definition. A prescheme X is catenary if whenever $Z \subseteq Z'$ are irreducible closed subsets of X, then every maximal chain

$$Z = Z_0 < Z_1 < \cdots < Z_n = Z'$$

of irreducible closed subsets between Z and Z' has the same length n.

Corollary 7.2. A locally noetherian prescheme X which admits a dualizing complex is catenary, and has finite Krull dimension.

Proof. Indeed, the existence of a codimension function d on X implies that X is catenary. On the other hand, since $R^{\cdot}$ is quasi-isomorphic to a bounded complex of injective sheaves, the function d is bounded in both directions, so X has finite Krull dimension.

Definition. Let X be a locally noetherian prescheme, and let $R^{\cdot}$ be a dualizing complex on X. We define the filtration $Z^{\cdot}$ associated to $R^{\cdot}$ by

$$Z^p = \{x \in X \mid d(x) \geq p\}$$

for each p. Then by virtue of the previous proposition and corollary, $Z^{\cdot}$ is a finite filtration of X, and each $x \in Z^p - Z^{p+1}$ is maximal.

Proposition 7.3. Let X be a locally noetherian prescheme, let $R^{\cdot}$ be a dualizing complex on X, and let $Z^{\cdot}$ be the associated filtration. Then $R^{\cdot}$ is Gorenstein with respect to $Z^{\cdot}$ (cf. [IV §3]). Moreover, for each $x \in X$, the injective hull $J(x)$ of $k(x)$ over $\mathcal{O}_x$ occurs exactly once as a direct summand of the (injective) complex $E(R^{\cdot})$.

Proof. Using condition (ii) of [IV 3.1] and [IV.1.F], it is sufficient to show that

$$H_x^i(R^\cdot) = \begin{cases} 0 & \text{for } i \neq d(x) \\ J(x) & \text{for } i = d(x). \end{cases}$$

But this follows from Proposition 6.1 above.

§8. Pointwise dualizing complexes and $f^{\#}$.

In this section we give a generalization of the notion of dualizing complex, which may exist on a locally noetherian prescheme of infinite Krull dimension. We apply this notion to show that $f^{\#}$ of a dualizing complex is dualizing, where f is a smooth morphism. Throughout this section, X will denote a locally noetherian prescheme.

Definition. A complex $R^{\cdot} \in D_c^+(X)$ has pointwise finite injective dimension (pfid) if for every $x \in X$, $R_x^{\cdot} \in D_c^+(\operatorname{Spec} \mathcal{O}_x)$ has finite injective dimension.

Proposition 8.1. Let $R^{\cdot} \in D_c^+(X)$ have pointwise finite injective dimension. Then conditions (i)-(iv) of Proposition 2.1 are equivalent.

Proof. The same as loc. cit. since one can check reflexivity pointwise.

Definition. A complex $R^{\cdot} \in D_c^+(X)$ with pfid is pointwise dualizing if it satisfies the equivalent conditions of the proposition.

Remarks. 1. Since fid implies pfid, we see that any dualizing complex is pointwise dualizing.

2. If $X = \operatorname{Spec} A$ where A is a local ring, then fid $\Longleftrightarrow$ pfid, so a complex is dualizing if and only if it is

pointwise dualizing.

3. We say a complex $F^\bullet \in D(X)$ is <u>pointwise bounded below</u> if for every $x \in X$, $F^\bullet_x \in D^+(\operatorname{Spec} \mathcal{O}_x)$, and we write $F^\bullet \in D^{p+}(X)$. Similarly we define $D^{p-}(X)$ and $D^{pb}(X)$. These are full subcategories of $D(X)$ (cf. [I §§4,5]). Now if $R^\bullet \in D_c^+(X)$ is pointwise dualizing, then the functor $\underline{\underline{R}}\ \underline{Hom}^\bullet(\cdot,R^\bullet)$ gives an anti-isomorphism of D^{p-} and D^{p+}, and an auto-anti-isomorphism of $D^{pb}(X)$.

4. If $R^\bullet$ is a pointwise dualizing complex on X, we can define as in §7 the associated codimension function $d: X \longrightarrow \mathbb{Z}$, and the associated filtration $Z^\bullet$. Proposition 7.1 holds in this case, so the filtration $Z^\bullet$ satisfies the conditions of [IV §3], in particular, it is separated. Thus we see as in Corollary 7.2 that X is catenary.

5. If $R^\bullet$ is pointwise dualizing on X, <u>and if $R^\bullet \in D_c^b(X)$</u>, then $R^\bullet$ is Gorenstein with respect to the filtration $Z^\bullet$ (cf. Proposition 7.3). The boundedness hypothesis is necessary for [IV 3.1].

<u>Example</u>. If X is a regular prescheme, then $\mathcal{O}_X$ is a pointwise dualizing complex for X. Indeed, for each $x \in X$,

Spec $\mathcal{O}_x$ is a regular prescheme of finite Krull dimension, so we apply Example 2.2 above.

<u>Proposition 8.2</u>. Let $R^\cdot \in D_c^+(X)$ be pointwise dualizing, and assume that X is noetherian and of finite Krull dimension. Then $R^\cdot$ has finite injective dimension and so is dualizing (by Proposition 2.3).

<u>Proof</u>. Since X is noetherian and of finite Krull dimension, the codimension function d associated to $R^\cdot$ is bounded. But we saw in the proof of Proposition 3.4 that for every $x \in X$, and every module M of finite type over $\mathcal{O}_x$,

$$\mathrm{Ext}^i_{\mathcal{O}_x}(M, R^\cdot_x) = 0 \qquad \text{for } i \notin [d(x)-\dim M, d(x)].$$

Thus in particular for every coherent sheaf F on X,

$$\underline{\mathrm{Ext}}^i_{\mathcal{O}_X}(F,R^\cdot) = 0 \qquad \text{for } i > \sup_{x\in X} d(x) ,$$

and so $R^\cdot$ has finite injective dimension [II 7.20 $(\underline{iii})_c$].

<u>Theorem 8.3</u>. Let $f\colon X \longrightarrow Y$ be a smooth morphism of locally noetherian preschemes, and let $R^\cdot \in D_c^+(Y)$ be pointwise dualizing. Then $f^\#(R^\cdot)$ (cf. [III §2]) is pointwise dualizing on X. If furthermore Y is noetherian, f of finite type, and $R^\cdot$ dualizing, then $f^\#(R^\cdot)$ is also dualizing.

Proof. For the first statement, the question is local on X and Y, so we reduce immediately to the case Y = Spec A, with A a local ring, and X of finite type, affine over Y. Then X admits a closed immersion into a suitable affine space $\mathbb{A}^n_Y$, so using [III 8.4] and Proposition 2.4 (which is valid also for "pointwise dualizing"), we reduce to the case $X = \mathbb{A}^n_Y$.

Case 1. A is a regular local ring. Then $R^\cdot$ is dualizing, since A is local, and by uniqueness (Theorem 3.1) we may assume, after shifting if necessary, that $R^\cdot \cong A$. Then $f^*(R^\cdot) \cong \omega_{X/Y} = \Omega^n_{X/Y}$, which is locally free of rank one, and hence pointwise dualizing (Example 2.2, since X is regular if Y is).

Case 2. A is a quotient of a regular local ring A'. Let Y' = Spec A'. Then X is obtained from the morphism $\mathbb{A}^n_{Y'} \longrightarrow Y'$ by the base extension $Y \longrightarrow Y'$, and we reduce to the previous case by means of [III.6.4], Proposition 2.4, the uniqueness (Theorem 3.1) on Y.

General case. A is an arbitrary noetherian local ring. Let $\hat{A}$ be the completion of A, and let Y' = Spec $\hat{A}$. We make the base extension $u: Y' \longrightarrow Y$, and then $u^*(R^\cdot)$ is dualizing on Y' by Corollary 3.5. But $\hat{A}$ is a quotient of a regular local ring by Cohen's structure theorem [14,(31.1)], so $f'^{*}(u^{*}(R^{\cdot}))$ is dualizing on $X' = X \times_Y Y'$, by Case 2. Now for any point $x \in X$, the completion $\hat{\mathcal{O}}_x$ of its local ring is also the completion of the ring $\mathcal{O}_x \otimes_A \hat{A}$, which is the semilocal ring of the points of X' lying over x, and so applying Corollary 3.5 once in each direction, and [III 2.1] we see that $f^{\#}(R^{\cdot})$ is pointwise dualizing on X.

Now if Y is noetherian, f of finite type, and $R^{\cdot}$ dualizing, then Y has finite Krull dimension by Corollary 7.2, hence X has finite Krull dimension, and so $f^{\#}(R^{\cdot})$ is dualizing by Proposition 8.2. q.e.d.

Corollary 8.4. Under the hypotheses of the theorem, let d_Y and d_X be the codimension functions associated with the pointwise dualizing complexes $R^{\cdot}$ and $f^{\#}(R^{\cdot})$, respectively. Then for each $x \in X$,

$$d_X(x) = d_Y(y) + \text{tr.d. } k(x)/k(y)$$

where $y = f(x)$.

Proof. Chasing through the reductions above (and noting that this formula holds trivially for a finite morphism and $f^{\flat}(R^{\cdot})$), we reduce to the case $Y = \operatorname{Spec} A$, A a regular local ring, $R^\cdot = A$, and $X = \mathbb{A}^n_Y$. Then $f^\#(R^\cdot) = \omega_{X/Y}$, which is locally free of rank one, so for any $x \in X$ we have $d_X(x) = \dim \mathcal{O}_x$, and for any $y \in Y$, $d_Y(y) = \dim \mathcal{O}_y$. [LC §4]. Our formula is then the usual dimension formula for a local homomorphism of local rings (cf. [EGA 0_{IV} 17.3.3]).

Remark. Using Proposition 2.4, the result of the theorem can be extended to show that for any embeddable morphism f of noetherian preschemes [III §8], $f^!$ takes dualizing complexes to dualizing complexes. The formula of the Corollary also extends to this case.

Proposition 8.5. Let $f: X \longrightarrow Y$ be an embeddable morphism of locally noetherian preschemes, let $R^\cdot \in D_c^+(Y)$ be a dualizing complex on Y, and let D denote the functor $\underline{\underline{R}}\ \underline{Hom}^\cdot_Y(\cdot, R^\cdot)$ or $\underline{\underline{R}}\ \underline{Hom}^\cdot_X(\cdot, f^! R^\cdot)$ (the one meant will be clear from the context). Then there is a functorial isomorphism

$$f^!(F^\cdot) \cong D(\underline{\underline{L}}f^*(D(F^\cdot)))$$

for all $F^\cdot \in D_c^+(Y)$.

Proof. Apply [III 8.8, 7)] to $D(F^\cdot)$ and $R^\cdot$.

<u>Corollary 8.6</u>. Under the hypotheses of the proposition, assume furthermore that f is flat, of finite type, and that Y is noetherian. Then there is a functorial isomorphism

$$f^{!}(F^{\bullet} \underline{\underline{\otimes}} G^{\bullet}) \xrightarrow{\sim} f^{*}(F^{\bullet}) \underline{\underline{\otimes}} f^{!}(G^{\bullet})$$

for $F^{\bullet} \in D_c^{+}(Y)$ and $G^{\bullet} \in D_c^{b}(Y)_{fTd}$.

<u>Proof</u>. Note first that Theorem 8.3 and the Remark following apply to show that $f^{!}(R^{\bullet})$ is dualizing on X. The result now follows from the Proposition, using [II 5.8] and Proposition 2.6 above.

§9. Gorenstein preschemes.

Theorem 9.1. Let A be a noetherian local ring. Then the following conditions are equivalent:

(i) A is a dualizing complex for itself.

(ii) A has a finite injective resolution.

(iii) A is Cohen-Macaulay-1 (i.e., A is Cohen-Macaulay, and whenever $x_1,\cdots,x_n$ is a maximal A-sequence, then the ideal $(x_1,\cdots,x_n)$ is irreducible.)

(iv) There is an integer d such that

$$\mathrm{Ext}^i(k,A) = \begin{cases} 0 & \text{for } i \neq d \\ k & \text{for } i = d \end{cases}$$

where k is the residue field of A.

(v) A is Gorenstein with respect to a suitable filtration $Z^\cdot$ of Spec A (cf. [IV §3]).

(vi) There is an integer d such that

$$H_x^i(A) \begin{cases} = 0 & \text{for } i \neq d \\ \text{is injective for} & i = d \end{cases}$$

where x is the closed point of Spec A.

<u>Proofs</u>. (i) $\Longrightarrow$ (ii) By definition of dualizing complex.

(ii) $\Longleftrightarrow$ (iii) $\Longleftrightarrow$ (iv) Proved by Bass [2, Theorem 4.1]. This article, incidentally, is a good summary of all previous occurrences of Gorenstein rings in the literature.

(iv) $\Longrightarrow$ (i) This is Proposition 3.4 above.

(i) $\Longrightarrow$ (v) This is Proposition 7.3 above.

(v) $\Longrightarrow$ (vi) By [IV 3.4], $E(A)$ is an injective resolution of A, and furthermore, for each $p \in \mathbb{Z}$, $E^p(A)$ is a direct sum of constant (injective) sheaves on subspaces $\{y\}^-$ of Spec A with $y \in Z^p - Z^{p+1}$. We can calculate $H^i_x(A)$ as $H^i(\Gamma_x(E(A)))$. But $\Gamma_x(E(A))$ by what we have just seen, consists of a single injective sheaf in some degree (say d), since x is a minimal point of Spec A.

(vi) $\Longrightarrow$ (iii) First note by [LC 6.4, part 4] that $H^n_x(A) \neq 0$, where $n = \dim A$, so $n = d$. (The hypothesis in loc. cit. that A is a quotient of a regular local ring can be circumvented by passing to the completion [LC 5.9].) Therefore A is Cohen-Macaulay [LC 3.10]. Let $(x_1, \cdots, x_n)$ be a maximal A-sequence, and let $\mathfrak{q} = (x_1, \cdots, x_n)$. Then

$H^n_x(A)$ is an essential extension of $A/\mathfrak{q}$ [11, proof of Prop. 2], and so is irreducible, since $A/\mathfrak{q}$ is a simple A-module, and hence irreducible. We conclude that $H^n_x(A)$ is an irreducible injective A-module with support at x, and so is an injective hull of k [II 7.4].

Now for any A-module M,

$$\operatorname{Hom}(k,M) = \operatorname{Hom}(k, \Gamma_x(M)).$$

Furthermore, the functor Γ_x takes injectives into injectives, so we have a spectral sequence of derived functors, which degenerates in this case to give

$$\operatorname{Ext}^i(k,A) = \operatorname{Hom}(k, H^n_x(A)) = \begin{cases} 0 & i \neq n \\ k & i = n \end{cases} .$$

<u>Definition</u>. A local noetherian ring satisfying the equivalent conditions of the theorem is called a (local) <u>Gorenstein ring</u>.

<u>Remark</u>. For a local ring, regular $\Longrightarrow$ complete intersection $\Longrightarrow$ Gorenstein $\Longrightarrow$ Cohen-Macaulay, and these implications are all strict [2].

<u>Definition</u>. A prescheme is <u>Gorenstein</u> if all of its local rings are Gorenstein local rings.

<u>Corollary 9.2</u>. A localization of a Gorenstein local ring is Gorenstein.

<u>Proof</u>. Follows from condition (i) of the Theorem, and Corollary 2.3 above.

<u>Proposition 9.3</u>. Let $f: X \longrightarrow \text{Spec } k$ be an embeddable morphism, where k is a field. Then X is Gorenstein if and only if $f^!(k)$ is isomorphic, in $D^+(X)$, to an invertible sheaf on X.

<u>Proof</u>. Indeed, by the remark at the end of the last section, $f^!(k)$ is a dualizing complex on X. But X is Gorenstein if and only if $\mathcal{O}_X$ is a dualizing complex on X. So the result follows from the uniqueness of the dualizing complex (Theorem 3.1 above).

<u>Corollary 9.4</u>. Let X be a Gorenstein prescheme of finite type over a field k, and let $k \subseteq K$ be a field extension. Then $X \otimes_k K$ is also Gorenstein.

<u>Proof</u>. The question is local on X, and $f^!$ is compatible with flat base extension [III 8.7, 5)].

<u>Corollary 9.5</u>. Let K,L be extension fields of a field k, and assume that one of them is finitely generated. Then $K \otimes_k L$ is a Gorenstein ring (i.e., every localization of it is a local Gorenstein ring).

<u>Proof</u>. Similar to the above. One can also give a direct proof, by induction on the number of generators. This generalizes [EGA IV 6.7.1.1] which says that $K \otimes L$ is Cohen-Macaulay.

<u>Proposition 9.6</u>. Let $f: X \longrightarrow Y$ be a flat morphism of locally noetherian preschemes, with Y Gorenstein (but no finiteness assumption on f). Then X is Gorenstein if and only if all the fibres X_y, for $y \in Y$, are Gorenstein preschemes.

<u>Proof</u>. The question is local, so we reduce to the following: let $A \longrightarrow B$ be a local homomorphism of local rings, with A Gorenstein, and B flat over A. Then B is Gorenstein if and only if $B/\mathfrak{m}_A B$ is Gorenstein.

To prove this, we use condition (iv) of Theorem 9.1. By flatness,

$$\operatorname{Ext}^i_B(B/\mathfrak{m}_A B, B) = \begin{cases} 0 & \text{for } i \neq d \\ B/\mathfrak{m}_A B & \text{for } i = d \end{cases}$$

for a suitable d. Thus the spectral sequence of Ext's for the change of ring from B to $B/\mathfrak{m}_A B$ degenerates, and we have

$$\mathrm{Ext}^i_B(k_B, B) \cong \mathrm{Ext}^{i-d}_{B/\mathfrak{m}_A B}(k_B, B/\mathfrak{m}_A B)$$

for each i, whence the result.

Exercise 9.7. Let $f: X \longrightarrow Y$ be a flat morphism of finite type of locally noetherian preschemes. Show that $f^!(\mathcal{O}_Y)$ (which is defined locally on X) is isomorphic in $D^+(X)$ to an invertible sheaf if and only if all the fibres X_y of f, for $y \in Y$, are Gorenstein preschemes. Also show that $f^!(\mathcal{O}_Y)$ has a unique non-zero cohomology sheaf, which is flat over Y, if and only if all the fibres of f are Cohen-Macaulay preschemes. Such morphisms are called Gorenstein morphisms (resp. Cohen-Macaulay morphisms).

Compare [EGA IV 6.3 and 6.7] for statements analogous to Corollary 9.5 and Propositinn 9.6 for Cohen-Macaulay.

§10. Existence of Dualizing Complexes.

We now draw together the experience gained in this chapter to make some remarks about the existence of a dualizing complex on a locally noetherian prescheme X. (We leave the analogous discussion for pointwise dualizing complexes to the reader.)

Sufficient conditions. 1. If X is Gorenstein and of finite Krull dimension, then $\mathcal{O}_X$ is a dualizing complex for X. In particular, any regular prescheme of finite Krull dimension has a dualizing complex.

2. If $f: X \longrightarrow Y$ is a morphism of finite type, with Y noetherian, and if Y admits a dualizing complex $R^{\cdot}$, we will see in the next chapter that X admits a dualizing complex $f^{!}(R^{\cdot})$. (If f is embeddable this result follows from §8 above.) Thus any noetherian prescheme of finite type over a Gorenstein prescheme of finite Krull dimension admits a dualizing complex.

3. In particular, any prescheme of finite type over a field k admits a dualizing complex.

4. If X is the spectrum of a complete local ring A, then X admits a dualizing complex. Indeed, A is a quotient of a regular local ring by the Cohen structure theorem [14, (31.1)].

Necessary conditions. 1. If X admits a dualizing complex, then X has finite Krull dimension (Corollary 7.2).

2. If X admits a dualizing complex, then X is catenary (in fact universally catenary since any prescheme of finite type over X admits a dualizing complex, at least locally).

3. More precisely, X admits a codimension function d.

4. If A is a local noetherian ring admitting a dualizing complex, and if $\mathfrak{p} \subseteq A$ is a prime ideal, then $\hat{A} \otimes_A k(\mathfrak{p})$ is a Gorenstein ring. This follows from the following

Proposition 10.1. If A is a local noetherian domain, with quotient field K, which admits a dualizing complex, then $\hat{A} \otimes_A K$ is a Gorenstein ring.

Proof. Let $R^{\cdot}$ be a dualizing complex on A. Then $R^{\cdot} \otimes_A \hat{A}$ is dualizing on $\hat{A}$ by Corollary 3.5, and so by localization, $(R^{\cdot}\otimes_A\hat{A})\otimes_A K$ is dualizing on $\hat{A} \otimes_A K$. But $R^{\cdot}\otimes_A K$ is dualizing on K, so by Theorem 3.1 we may assume $R^{\cdot}\otimes_A K \cong K$ after translating if necessary. But $(R^{\cdot}\otimes_A\hat{A})\otimes_A K \cong (R^{\cdot}\otimes_A K)\otimes_K(\hat{A}\otimes_A K)$ so $\hat{A} \otimes_A K$ is a dualizing complex for itself, and we are done (Theorem 9.1 (i)).

<u>Example</u>. There are local noetherian domains of dimension 3 which are not catenary, and local noetherian domains of dimension 2 which are not universally catenary, and which therefore have no dualizing complexes [14, Appendix A1, Ex. 2].

<u>Problems</u>. 1. We do now know if $\hat{A} \otimes_A K$ is a Gorenstein ring, even if A is a local domain of dimension 1, hence we do not know whether every local domain of dimension 1 admits a dualizing complex.

2. If X is noetherian and of finite Krull dimension, and if it admits a dualizing complex locally, and if it admits a codimension function d, we do not know whether X admits a dualizing complex globally.

CHAPTER VI. RESIDUAL COMPLEXES

§0. Introduction.

In this chapter we return to the problem of constructing a functor $f^!$ for a morphism of finite type, which should reduce to $f^\flat$ for a finite morphism, and $f^\#$ for a smooth morphism. We ran into difficulty earlier [III §8] because the derived category is not a local object — one cannot glue together elements of the derived category given locally. Now we overcome that difficulty in a special case by using residual complexes. The residual complex is a very special complex of quasi-coherent injective sheaves (see definition below) which is almost unique (it was called "residue complex" in the Edinburgh Congress talk [9]).

Modulo some technicalities arising from possibly infinite Krull dimension, this is how they work: To each dualizing complex $R^\cdot \in D_c^+(X)$ is associated functorially a residual complex $K^\cdot = E(R^\cdot)$, and $Q(K^\cdot)$ (which is the image of the complex $K^\cdot$ in $D_c^+(X)$) is isomorphic to $R^\cdot$ in $D_c^+(X)$. Thus we will define $f^!$ locally for a residual complex $K^\cdot$ by

$$f^!(K^\cdot) = E(f^!(Q(K^\cdot))),$$

where $f^!$ is the one we know for embeddable morphisms. Then since $f^!(K^\cdot)$ is an actual complex defined locally, we can glue to get a global $f^!(K^\cdot)$. We will give the statement and proof of the existence of $f^!$ in some detail, since this is an important step towards the general duality theorem. Later, after proving the duality theorem, we will pull ourselves up by our bootstraps to obtain a definition of $f^!$ for objects in the derived category $D_c^+(X)$.

Once having $f^!$, we will define a trace map for residual complexes, which will be a map of graded sheaves

$$Tr_f\colon \; f_* f^! K^\cdot \longrightarrow K^\cdot.$$

We will prove in the next chapter that if f is a proper morphism, then Tr_f is a map of complexes (Residue Theorem).

§1. Residual Complexes.

Throughout this section, X will denote a locally noetherian prescheme. If x is a point of X, and if I is an injective hull of $k(x)$ over the local ring $\mathcal{O}_x$, we will denote by $J(x)$ the quasi-coherent injective $\mathcal{O}_X$-module, which is the constant sheaf I on $\{x\}^-$, and 0 elsewhere (notation of [II §7]).

Definition. A residual complex on X is a complex $K^\cdot$ of quasi-coherent injective $\mathcal{O}_X$-modules, bounded below, with coherent cohomology sheaves, and such that there is an isomorphism

$$\sum_{p\in\mathbb{Z}} K^p \cong \sum_{x\in X} J(x) \quad .$$

Example. If X is a regular prescheme, then the Cousin complex of the structure sheaf $\mathcal{O}_X$ is a residual complex for X (see example at end of [IV §2]).

Proposition 1.1. a) If $K^\cdot$ is a residual complex on X, then its image $Q(K^\cdot) \in D_c^+(X)$ is a pointwise dualizing complex.

b) If $R^\cdot \in D_c^+(X)$ is a pointwise dualizing complex on X, then $E(R^\cdot)$ is a residual complex on X. (Here E is the notation of [IV §3], with respect to the filtration $Z^\cdot$ associated with $R^\cdot$

(cf. [V §7] and [V §8, Remark 4]).)

c) If X admits a residual complex (or pointwise dualizing complex) with bounded cohomology, then there is a functorial isomorphism

$$\varphi\colon\ R^{\cdot} \xrightarrow{\ \sim\ } QE(R^{\cdot})$$

for pointwise dualizing complexes. Hence the functor

$$E\colon\ \text{Ptwdual}\ (X) \longrightarrow \text{Res}\ (X)$$

is an equivalence of the category of pointwise dualizing complexes of $D_c^+(X)$ and the category of residual complexes (and morphisms of complexes). Its inverse is Q.

Proof. a) The question is local, so we may assume $X = \operatorname{Spec} A$, where A is a noetherian local ring. By [V.3.4] we have only to check that there is an integer d such that

$$\operatorname{Ext}^i(k,K^{\cdot}) = \begin{cases} 0 & \text{for}\quad i \neq d \\ k & \text{for}\quad i = d \end{cases}$$

where k is the residue field of A. Since $k = k(x_o)$, where x_o is the closed point of X, we have

$$\operatorname{Hom}\ (k,\ J(x_o)) = k$$

$$\operatorname{Hom}\ (k,\ J(x)) = 0 \qquad \text{for}\quad x \neq x_o.$$

The result now follows from the definition of a residual complex.

b) By the pointwise convergence of the spectral sequence [IV.1.G] we see that $H^i(R^\cdot) = H^i(E(R^\cdot))$ for all i, and hence $E(R^\cdot)$ is a complex which is bounded below and has coherent cohomology. By [IV.1.F] the question of whether it is a residual complex is local, so we may assume X is the spectrum of a local ring, in which case our result is [V.7.3].

c) This follows from [V, §8, Remark 5] and [IV.3.4].

Remarks. 1. In particular, if X admits a dualizing complex (and hence has finite Krull dimension), the functor

$$E: \text{Dual}(X) \longrightarrow \text{Res}(X)$$

is an equivalence of categories, with inverse Q.

2. I expect that the statement c) is false without the boundedness assumption. That is, there may be two non-isomorphic pointwise dualizing complexes $R^\cdot$ and $R'^\cdot$ such that $E(R^\cdot) = E(R'^\cdot)$. In particular, there may not be a uniqueness theorem similar to [V.3.1] for pointwise dualizing complexes.

3. It follows from the proposition that X admits a residual complex if and only if it admits a pointwise dualizing complex, so the remarks of [V §10] apply.

<u>Exercise</u>. Show that there is a uniqueness theorem analogous to [V.3.1] for residual complexes, i.e., two residual complexes can differ only by shifting degrees and tensoring with an invertible sheaf. The touchy point is to show that if $K^{\cdot}, K'^{\cdot}$ are residual complexes, then $Hom^{\cdot}(K^{\cdot}, K'^{\cdot})$ is a complex with coherent cohomology!

We now give two technical results which will be used in the following sections.

<u>Lemma 1.2</u>. Let $K^{\cdot}$ and $K'^{\cdot}$ be residual complexes on X. Then to give an isomorphism $\psi: K^{\cdot} \longrightarrow K'^{\cdot}$ is equivalent to giving, for each $x \in X$, an isomorphism

$$\psi_x: Q(K_x^{\cdot}) \longrightarrow Q(K_x'^{\cdot})$$

in $D_c^+(\mathrm{Spec}\ \mathcal{O}_x)$, such that whenever $x \longrightarrow y$ is a specialization, then ψ_x is obtained from ψ_y by localization.

<u>Proof</u>. Clearly ψ gives rise to a system $(\psi_x)_{x\in X}$ as described. Conversely, given the isomorphisms ψ_x, we first note that by c) of the Proposition above, ψ_x comes from a unique isomorphism

$$\overline{\psi}_x : K^{\cdot}_x \longrightarrow K'^{\cdot}_x$$

of the actual residual complexes, and these $\overline{\psi}_x$ are compatible with localization. Already we deduce that the codimension functions d and d' associated to $K^{\cdot}$ and $K'^{\cdot}$ are the same. And since for $d(x) = d(y)$ and $x \neq y$ there are no non-zero maps of $J(x)$ into $J(y)$, our system of isomorphisms $\overline{\psi}_x$ gives rise to, and is determined by, a collection of isomorphisms

$$\varphi_x : I(x) \longrightarrow I'(x)$$

for each $x \in X$, where $I(x)$ (resp. $I'(x)$) is the (unique) subsheaf of $K^{\cdot}$ (resp. $K'^{\cdot}$) isomorphic to $J(x)$, such that for an immediate specialization $x \longrightarrow y$, φ_x and φ_y are compatible with the boundary maps of the complexes $K^{\cdot}$ and $K'^{\cdot}$. But to give such a collection of isomorphisms (φ_x) is to give an isomorphism $\psi: K^{\cdot} \longrightarrow K'^{\cdot}$, so we are done.

<u>Lemma 1.3</u> (Glueing Lemma). Let (U_ν) be an open cover of X, and let $K^{\cdot}_\nu$ be a residual complex on U_ν, for each ν. Suppose furthermore that for each pair of indices μ, ν we are given an isomorphism

$$\alpha_{\mu\nu}\colon K^{\cdot}_{\nu} \longrightarrow K^{\cdot}_{\mu}$$

of the restrictions of these complexes to $U_{\mu\nu} = U_{\mu} \cap U_{\nu}$, such that for each triple μ,ν,λ,

$$\alpha_{\mu\nu}\alpha_{\nu\lambda}\alpha_{\lambda\mu} = 1$$

on $U_{\mu\nu\lambda}$. Suppose finally that the lower bound of the complexes $K^{\cdot}_{\nu}$ can be chosen uniformly for all ν. Then there exists a unique residual complex $K^{\cdot}$ on X, together with isomorphisms

$$\beta_{\nu}\colon K^{\cdot}|_{U_{\nu}} \longrightarrow K^{\cdot}_{\nu}$$

for each ν, which are compatible with $\alpha_{\mu\nu}$ on $U_{\mu\nu}$.

Proof. This is the usual situation for glueing, so it is clear that we can glue the complexes $K^{\cdot}_{\nu}$ into a global complex $K^{\cdot}$ on X, which is bounded below. It is also clear that $K^{\cdot}$ is a complex of quasi-coherent injective $\mathcal{O}_X$-modules [II.7.16], and has coherent cohomology. Finally since the $J(x)$ are constant sheaves on irreducible subsets of X, we can glue together local isomorphisms to obtain a global isomorphism

$$\sum_{p\in\mathbb{Z}} K^p \cong \sum_{x\in X} J(x).$$

Hence $K^{\cdot}$ is a residual complex.

<u>Remark</u>. This lemma has the interesting consequence that one can glue dualizing complexes given locally in $D_c^+(U_\nu)$ into a global dualizing complex in $D_c^+(X)$, whereas one cannot glue arbitrary objects of the derived category given locally (cf. remarks at end of [III §8]).

§2. Functorial Properties.

In this section we define a notion of $f^!$ for residual complexes in case f is a finite or smooth morphism. To avoid confusion, we introduce new, temporary notation f^y and f^z for these functors. The letters y,z are arbitrary symbols; they have nothing to do with schemes Y,Z, or with points thereof. We give four isomorphisms I-IV between these functors, and seven compatibilities (i)-(vii) which will be referred to in the next two sections.

If $f: X \longrightarrow Y$ is a finite morphism of locally noetherian preschemes, we define a functor

$$f^y: \text{Res}(Y) \longrightarrow \text{Res}(X)$$

on the category of residual complexes by

$$f^y(K^\cdot) = Ef^\flat Q(K^\cdot).$$

Note by Proposition 1.1a that $Q(K^\cdot)$ is a pointwise dualizing complex; by [V.2.4] $f^\flat Q(K^\cdot)$ is pointwise dualizing on X, and by Proposition 1.1b, $Ef^\flat Q(K^\cdot)$ is a residual complex on X.

If $f: X \longrightarrow Y$ and $g: Y \longrightarrow Z$ are two finite morphisms of locally noetherian preschemes, we define an isomorphism

$$(gf)^{\nabla} \xrightarrow{\sim} f^{\nabla}g^{\nabla} \qquad (I)$$

of functors from Res(Z) to Res(X) as follows. By Lemma 1.2 it will be enough to define this isomorphism in the scheme $\mathrm{Spec}(\mathcal{O}_x)$ for each $x \in X$, in a manner compatible with localization. A fortiori, it is enough to define the isomorphism after making a base extension $\mathrm{Spec}(\mathcal{O}_z) \longrightarrow Z$, for each point $z \in Z$, and thus we reduce to the case where Z (and hence also X and Y) are noetherian of finite Krull dimension. In that case pointwise dualizing complexes are dualizing, and we have a functorial isomorphism $\varphi: 1 \xrightarrow{\sim} QE$ on the category of dualizing complexes on Y (Proposition 1.1c). Now for a residual complex $K^{\cdot}$ on Z we define our isomorphism as follows:

$$(gf)^{\nabla}K^{\cdot} \underset{\text{def.}}{=} E(gf)^{\flat}QK^{\cdot} \underset{[\text{III } 6.2]}{\cong} Ef^{\flat}g^{\flat}QK^{\cdot} \underset{\varphi}{\cong} Ef^{\flat}QE\,g^{\flat}QK \underset{\text{def.}}{=} f^{\nabla}g^{\nabla}(K^{\cdot}).$$

We will use this same technique of reduction to the case of finite Krull dimension without explicit mention below. It enables us to carry over isomorphisms defined for dualizing complexes to residual complexes.

For a composition of three finite morphisms, there is the usual commutative diagram (referred to as (i)) of the isomorphisms (I).

For a smooth morphism $f: X \longrightarrow Y$ of locally noetherian preschemes we define a functor

$$f^Z: \text{Res}(Y) \longrightarrow \text{Res}(X)$$

by

$$f^Z(K^\cdot) = Ef^{\#}Q(K^\cdot) .$$

This takes residual complexes into residual complexes by virtue of Proposition 1.1a,b, and [V.8.3].

For a composition of two smooth morphisms f,g, we define an isomorphism

$$(gf)^Z \xrightarrow{\sim} f^Z g^Z \qquad \text{(II)}$$

using the above reduction to the case of finite Krull dimension, and carrying over the isomorphism [III 2.2].

For a composition of three smooth morphisms, there is a compatibility (referred to as (ii)) of the isomorphisms (II), which follows from the compatibility of [III 2.2].

There are two other isomorphisms, expressing compatibilities between the functors f^Y and g^Z: the Cartesian square, and the residue isomorphism. For the first, we suppose given a Cartesian diagram as shown (i.e., $W = X \times_Z Y$) with f (and hence k) a finite morphism, and g (and hence h) a smooth morphism. In that case there is an isomorphism

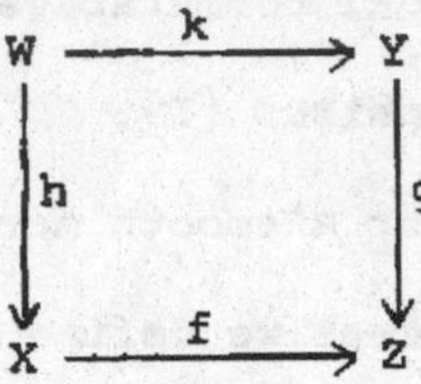

$$h^Z f^Y \xrightarrow{\sim} k^Y g^Z \qquad \text{(III)}$$

obtained as above from the isomorphism of [III 6.4].

If we have another Cartesian diagram as shown, with r also a finite morphism, then there is a commutative diagram of isomorphisms

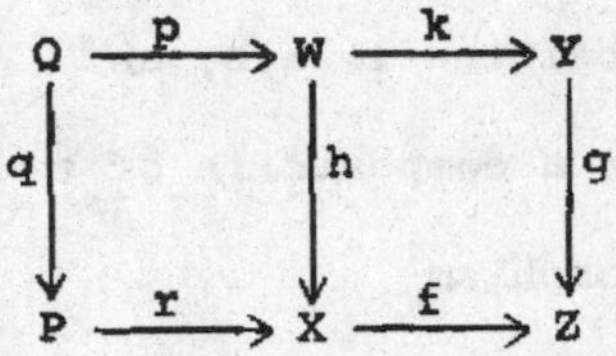

$$\begin{array}{ccccc} q^Z(fr)^Y & \xrightarrow{\text{III}} & (kp)^Y g^Z & & \\ \downarrow \text{I} & & & \searrow \text{I} & \\ q^Z r^Y f^Y & \xrightarrow{\text{III}} & p^Y h^Z f^Y & \xrightarrow{\text{III}} & p^Y k^Y g^Z \end{array} \qquad \text{(iii)}$$

Also if we have another Cartesian diagram doubling the smooth side of the square, i.e., with r smooth, then there is a similar commutative diagram (iv) involving the isomorphisms (II) and (III). These two compatibilities follow from [III 6.4].

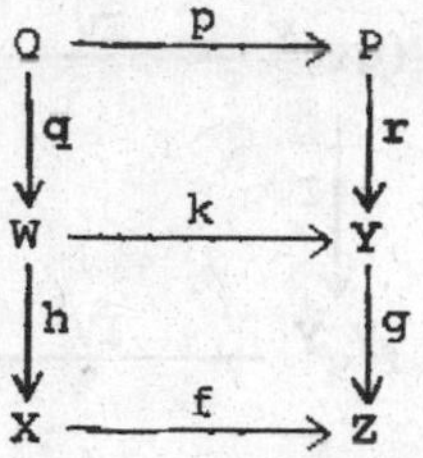

For the residue isomorphism, we suppose that we have a finite morphism f which is factored into pi, with i a closed immersion, and p smooth. Then there is an isomorphism

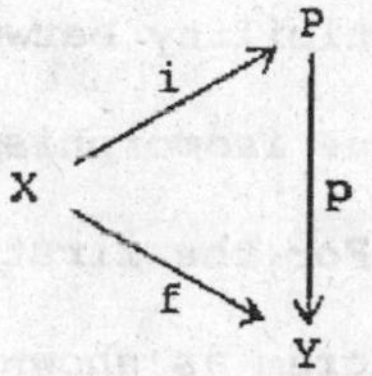

$$f^Y \xrightarrow{\sim} i^Y p^Z \qquad \text{(IV)}$$

obtained as above from that of [III 8.2].

If we have a diagram such as the one shown, with f,g finite, i,k closed immersions, and q,p smooth, then there is a commutative diagram

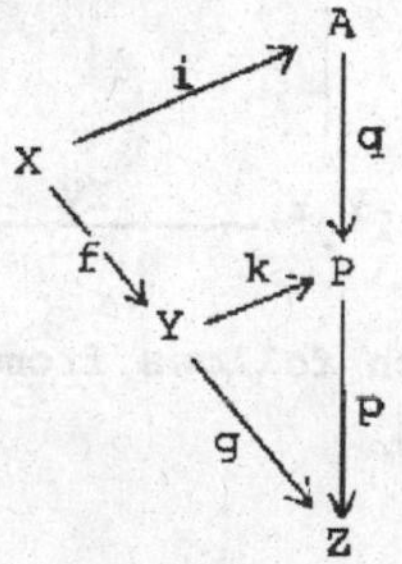

$$\begin{array}{ccccc} (gf)^Y & \xrightarrow{\text{IV}} & i^Y(pq)^Z & \xrightarrow{\text{II}} & i^Y q^Z p^Z \\ \Big\downarrow{\scriptstyle \text{I}} & & & & \Big\uparrow{\scriptstyle \text{IV}} \\ f^Y g^Y & \xrightarrow{\text{IV}} & f^Y k^Y p^Z & \xrightarrow{\text{I}} & (kf)^Y p^Z \end{array} \quad . \qquad \text{(v)}$$

This follows from [III 8.6b] applied to the triples (i,q,p) and (f,k,p).

There are also two commutative diagrams expressing compatibility between the square isomorphism (III) and the residue isomorphism (IV).

For the first, suppose one has a diagram as shown with $Q = Y \times_Z P$, i finite, j,k,ℓ closed immersions, and f,g smooth. Then there is a commutative diagram

$$\begin{array}{ccccc} & & Q & \xrightarrow{\ell} & P \\ & {\scriptstyle k}\nearrow & \Big\downarrow{\scriptstyle g} & & \Big\downarrow{\scriptstyle f} \\ X & \xrightarrow{i} & Y & \xrightarrow{j} & Z \end{array}$$

$$\begin{array}{ccccc} (ji)^Y & \xrightarrow{\text{IV}} & (\ell k)^Y f^Z & & \\ \Big\downarrow & & & \searrow{\scriptstyle \text{I}} & \\ i^Y j^Y & \xrightarrow{\text{IV}} & k^Y g^Z j^Y & \xrightarrow{\text{III}} & k^Y \ell^Y f^Z \end{array} \quad , \qquad \text{(vi)}$$

which follows from [III 8.6c].

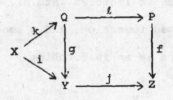

For the second, suppose there is a Cartesian diagram as shown ($W = X \times_Y Z$; $Q = P \times_Y Z$) with f,g finite, i,j closed immersions, and p,q,u,v,w smooth. Then there is a commutative diagram

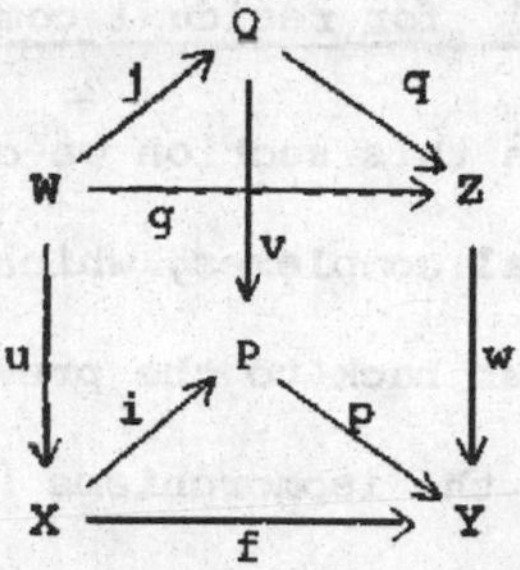

$$\begin{array}{ccccc} u^Z f^Y & \xrightarrow{\text{IV}} & u^Z i^Y p^Z & \xrightarrow{\text{III}} & j^Y v^Z p^Z \\ \Big| {\scriptstyle \text{III}} & & & & \searrow {\scriptstyle \text{II}} \\ g^Y w^Z & \xrightarrow{\text{IV}} & j^Y q^Z w^Z & \xrightarrow{\text{II}} & j^Y(wq)^Z = j^Y(pv)^Z \end{array} \qquad \text{(vii)}$$

This follows from [III 8.6a] modified as in [III 6.4].

§3. $f^!$ for residual complexes.

In this section we construct the functor $f^!$ for residual complexes, which we call f^Δ to avoid confusion. We refer back to the previous section for the notations f^Y, f^Z, the isomorphisms (I)-(IV), and the compatibilities (i)-(vii).

We will work in the category of locally noetherian preschemes, and we will consider only morphisms which are of finite type, and such that the dimensions of the fibres are bounded. It will be understood in the following that these conditions hold for all schemes and morphisms considered.

We are now in a position to state our theorem.

Theorem 3.1. There exists a theory of variance consisting of the data a)-d) below subject to the conditions VAR 1 - VAR 5. Furthermore, this theory is unique in the sense that given a second collection of such data a')-d') there is a unique isomorphism of the functors a) and a') compatible with the isomorphisms b)-d) and b')-d').

a) For every morphism $f: X \longrightarrow Y$, a functor

$$f^\Delta : \text{Res}(Y) \longrightarrow \text{Res}(X)$$

on the category of residual complexes.

b) For every pair of consecutive morphisms $f: X \longrightarrow Y$ and $g: Y \longrightarrow Z$, an isomorphism

$$c_{f,g}: (gf)^{\Delta} \xrightarrow{\sim} f^{\Delta}g^{\Delta} .$$

c) For every finite morphism f, an isomorphism

$$\psi_f: f^{\Delta} \xrightarrow{\sim} f^{Y} .$$

d) For every smooth morphism g, an isomorphism

$$\varphi_g: g^{\Delta} \xrightarrow{\sim} g^{Z} .$$

VAR 1). For any f, $c_{id,f} = c_{f,id} = id$, and if f,g,h are three consecutive morphisms, then there is a commutative diagram

$$\begin{array}{ccc} (hgf)^{\Delta} & \xrightarrow{c_{f,hg}} & f^{\Delta}(hg)^{\Delta} \\ \Big\downarrow c_{gf,h} & & \Big\downarrow c_{g,h} \\ (gf)^{\Delta}h^{\Delta} & \xrightarrow{c_{f,g}} & f^{\Delta}g^{\Delta}h^{\Delta} \end{array} .$$

(In other words, the f^{Δ} and $c_{f,g}$ define a "catégorie clivée normalisée" in the terminology of [SGA 1960-61, VI]).

VAR 2). If f,g are consecutive finite morphisms, then $c_{f,g}$ is compatible, via ψ_f and ψ_g, with the usual isomorphism (I) above.

VAR 3). If f,g are consecutive smooth morphisms, then $c_{f,g}$ is compatible, via φ_f and φ_g, with the usual isomorphism (II) above.

VAR 4). Given a Cartesian diagram as shown, with f,k finite and g,h smooth, the isomorphisms b) are compatible via c), d) with the isomorphism (III) of a square above, i.e., there is a commutative diagram

$$\begin{array}{ccc} W & \xrightarrow{k} & Y \\ \downarrow{\scriptstyle h} & & \downarrow{\scriptstyle g} \\ X & \xrightarrow{f} & Z \end{array}$$

$$\begin{array}{ccccc} (fh)^{\Delta} = (gk)^{\Delta} & & \xrightarrow{c_{k,g}} & & k^{\Delta}g^{\Delta} \\ \downarrow{\scriptstyle c_{h,f}} & & & & \downarrow{\scriptstyle \psi_k,\varphi_g} \\ h^{\Delta}f^{\Delta} & \xrightarrow{\varphi_h,\psi_f} & h^{Z}f^{Y} & \xrightarrow{III} & k^{Y}g^{Z} \end{array} .$$

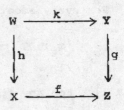

VAR 5). Given a diagram as shown, with i a closed immersion, f finite, and p smooth, we have a commutative diagram

$$\begin{array}{ccc} & & P \\ & \overset{i}{\nearrow} & \Big\downarrow p \\ X & \underset{f}{\searrow} & Y \end{array}$$

$$\begin{array}{ccc} f^{\Delta} = (pi)^{\Delta} & \xrightarrow{c_{i,p}} & i^{\Delta} p^{\Delta} \\ \Big\downarrow \psi_f & & \Big\downarrow \psi_i, \varphi_p \\ f^{Y} & \xrightarrow{IV} & i^{Y} p^{Z} \end{array} ,$$

i.e., $c_{i,p}$ is compatible with the residue isomorphism.

The proof of this theorem requires drawing a great many diagrams and checking their commutativity. We will therefore carry out in detail only a few of these verifications, by way of example, and will leave the others to the reader, marking them with the symbol (!) which indicates that he has some work to do at that point. The proof per se will follow after some definitions and lemmas.

<u>Definition</u>. Let $f: X \longrightarrow Y$ be a fixed morphism, and let $U \subseteq X$ be an open set. We define a <u>chart</u> on U to be the following collection of data:

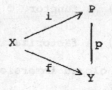

1) A functor f^a: Res(Y) $\longrightarrow$ Res(U).

2) A factorization $f|_U = pi$ where i is a closed immersion into a scheme P smooth over Y.

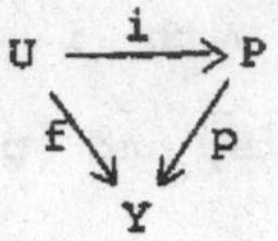

3) An isomorphism

$$\gamma_{i,p}\colon\ f^a \xrightarrow{\ \sim\ } i^y p^z \ .$$

<u>Definition</u>. If f^b, $f|_U = jq$, $q\colon Q \longrightarrow Y$, and $\gamma'_{j,q}$ is a second chart on the same open set U, a <u>permissible</u> isomorphism between the two is an isomorphism

$$\alpha\colon\ f^a \longrightarrow f^b$$

of functors, such that for every commutative diagram such as the one shown, with k a closed immersion and r,s smooth, there is a commutative diagram of isomorphisms

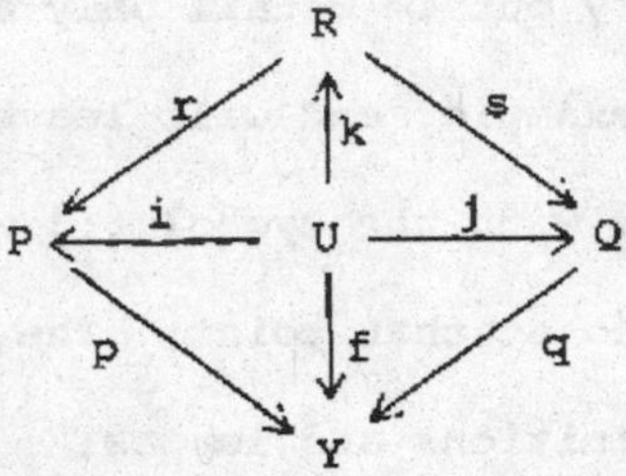

$$\begin{array}{ccccccc}
f^a & \xrightarrow{\gamma_{i,p}} & i^y p^z & \xrightarrow{\text{IV}} & k^y r^z p^z & & \\
\downarrow{\scriptstyle \alpha} & & & & & \searrow{\scriptstyle \text{II}} & \\
f^b & \xrightarrow{\gamma'_{j,q}} & j^y q^z & \xrightarrow{\text{IV}} & k^y s^z q^z & \xrightarrow{\text{II}} & k^y(qs)^z = k^y(pr)^z .
\end{array}$$

Lemma 3.2. Given two charts on an open set U, there exists a unique permissible isomorphism between them.

Proof. The uniqueness is clear, because one can take $R = P \times_Y Q$ and k the diagonal map $i \times j$. Then the isomorphism α is determined by the condition of the definition.

For the existence, let f^a, f^b, etc., be two charts as above, and let (R,k,r,s) and $(S,\ell,\cdots)$ be two diagrams as above. We must show that the isomorphisms α_R and α_S defined by the condition above are the same. By considering the third diagram $(R \times S,\ k \times \ell, \ldots)$ and comparing each to this one, we reduce to the case where S dominates R. In other words, we have a commutative diagram, such as the one shown, with k, ℓ closed immersions and r,s,t smooth morphisms. We must show that the following diagram is commutative:

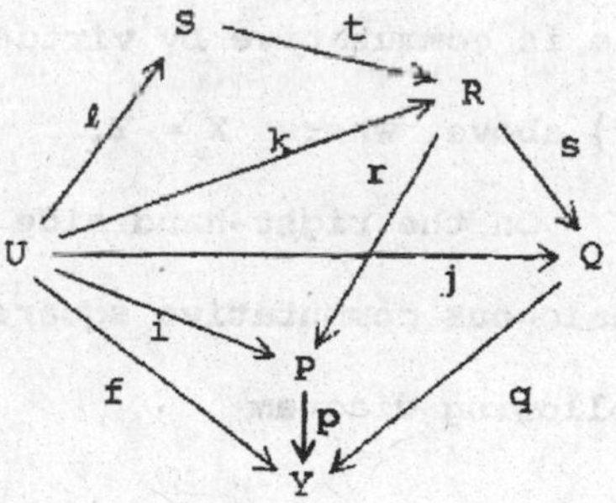

$$\begin{array}{ccccccccccccc}
f^a & \xrightarrow{\gamma_{i,p}} & i^Y p^Z & \xrightarrow{IV} & k^Y r^Z p^Z & \xrightarrow{II} & k^Y(pr)^Z = k^Y(qs)^Z & \xleftarrow{II} & k^Y s^Z q^Z & \xleftarrow{IV} & j^Y q^Z & \xleftarrow{\gamma'_{i,q}} & f^b \\
\downarrow \mathrm{id} & & & & & & & & & & & & \downarrow \mathrm{id} \\
f^a & \xrightarrow{\gamma_{i,p}} & i^Y p^Z & \xrightarrow{IV} & \ell^Y (rt)^Z p^Z & \xrightarrow{II} & \ell^Y(prt)^Z = \ell^Y(qst)^Z & \xleftarrow{II} & \ell^Y (st)^Z q^Z & \xleftarrow{IV} & j^Y q^Z & \xleftarrow{\gamma'_{i,q}} & f^b
\end{array}$$

We will check this one completely by filling in more arrows to make little commutative diagrams. At the left hand end of the diagram we fill in as follows:

$$\begin{array}{ccccc}
f^a & \xrightarrow{\gamma_{i,p}} & i^Y p^Z & \xrightarrow{IV} & k^Y r^Z p^Z \\
\Big\downarrow{\scriptstyle id} & & \Big\downarrow{\scriptstyle id} & & \Big\downarrow{\scriptstyle IV} \\
 & & & & \ell^Y t^Z r^Z p^Z \\
 & & & & \Big\downarrow{\scriptstyle II} \\
f^a & \xrightarrow{\gamma_{i,p}} & i^Y p^Z & \xrightarrow{IV} & \ell^Y (rt)^Z p^Z
\end{array} \quad .$$

The left-hand square is identically commutative. The right-hand one is commutative by virtue of a special case of compatibility (v) above, where $X = Y$.

On the right-hand side of our long diagram we fill in two analogous commutative squares. This leaves in the middle the following diagram

$$\begin{array}{ccccccc}
k^Y r^Z p^Z & \xrightarrow{II} & k^Y (pr)^Z & = & k^Y (qs)^Z & \xleftarrow{II} & k^Y s^Z q^Z \\
\Big\downarrow{\scriptstyle IV} & & \Big\downarrow{\scriptstyle IV} & & \Big\downarrow{\scriptstyle III} & & \Big\downarrow{\scriptstyle IV} \\
\ell^Y t^Z r^Z p^Z & \xrightarrow{II} & \ell^Y t^Z (pr)^Z & = & \ell^Y t^Z (qs)^Z & \xleftarrow{II} & \ell^Y t^Z s^Z q^Z \\
\Big\uparrow{\scriptstyle II} & & \Big\uparrow{\scriptstyle II} & & \Big\uparrow{\scriptstyle II} & & \Big\uparrow{\scriptstyle II} \\
\ell^Y (rt)^Z p^Z & \longrightarrow & \ell^Y (prt)^Z & = & \ell^Y (qst)^Z & \xleftarrow{II} & \ell^Y (ts)^Z q^Z
\end{array} \quad .$$

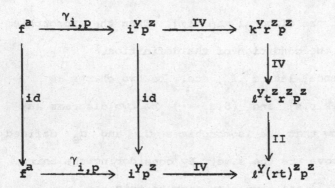

The middle squares are obviously commutative; the upper left and upper right are commutative because the isomorphism IV operates between U and R, and II operates between R and Y, so that the order doesn't matter; and the lower left and lower right are commutative by (ii) above.

q.e.d. lemma.

Lemma 3.3. The composition of permissible isomorphisms is permissible. The inverse of a permissible isomorphism is permissible.

Proof. The inverse is obvious. For the composition, let f^a and f^b be two charts as above, and let f^c, ℓ, r, R be a third. Using $P \times Q \times R$ to construct the unique permissible isomorphisms of f^a with f^b, f^b with f^a, and f^a with f^c, we see that the composition of the first two is the third.

Proof of theorem. First we prove the existence of the theory of variance.

a) Construction of the functor f^{Δ}. Let $f: X \longrightarrow Y$ be a morphism. Let $\mathcal{U} = (U_\nu)$ be a cover of X by open sets with charts on them. Note that charts exist locally: If $x \in X$ is a point, let V be a noetherian affine neighborhood of $f(x)$ in Y, and let U be an affine neighborhood of x in $f^{-1}(V)$, which is of finite type over V. Then U admits a closed immersion i into an affine space $\mathbb{A}^n_V$ for suitable n. Let $P = \mathbb{A}^n_V$, let $p: P \longrightarrow Y$ be the natural projection, let $f^a = i^Y p^z$, and $\gamma_{i,p} = id$. This gives a chart on U which is a neighborhood of the given point x.

If f^a, i, P, p, $\gamma_{i,p}$ is a chart on an open set U, and if $U' \subseteq U$ is a smaller open set, we define the notion of a <u>restriction</u> (not unique) of the chart to U', as follows: let $P' \subseteq P$ be an open set whose intersection with i(U) is i(U'). Then we take $f^a|_{U'}$, $i|_{U'}$, P', $p|_{P'}$, $\gamma_{i,p}|_{U'}$ as the restriction of the chart. Note that a restriction of permissible isomorphisms is permissible.

For each pair of indices μ,ν, choose restrictions of the charts on U_ν and U_μ to $U_{\mu\nu} = U_\nu \cap U_\mu$, and let $\alpha_{\mu\nu}$ be the unique permissible isomorphism between them. One sees immediately that the isomorphism of functors

$$\alpha_{\mu\nu}: f^\nu \longrightarrow f^\mu$$

thus defined on $U_{\mu\nu}$ is independent of the restrictions of the charts chosen. Furthermore it follows from the lemmas that on a triple intersection $U_{\mu\nu\lambda}$, these isomorphisms are compatible:

$$\alpha_{\mu\nu}\alpha_{\nu\lambda}\alpha_{\lambda\mu} = \mathrm{id}.$$

Thus, since we are dealing with functors on residual complexes, we can use Lemma 1.3, and glue the functors f^{ν} via the isomorphisms $\alpha_{\mu\nu}$ to obtain a functor

$$f^{\Delta}: \mathrm{Res}(Y) \longrightarrow \mathrm{Res}(X)$$

(together with isomorphisms $\beta_{\nu}: f^{\Delta}|_{U_{\nu}} \longrightarrow f^{\nu}$ for each ν compatible with the $\alpha_{\mu\nu}$) which is the one we want.

b) Construction of the isomorphisms $c_{f,g}$. Let $f: X \longrightarrow Y$ and $g: Y \longrightarrow Z$ be morphisms. It will be sufficient to construct $c_{f,g}$ locally, chart by chart, and show that it is compatible with the permissible isomorphisms of change of chart. For then we can glue. Thus we may assume that f and g are embeddable in smooth morphisms, and we may even assume that X is embeddable in an affine space over Y, since that is possible locally. Choose embeddings $i: X \longrightarrow \mathbb{A}^n_Y$ and $k: Y \longrightarrow P$, smooth over Z, and note that $\mathbb{A}^n_Y = \mathbb{A}^n_P \times_P Y$, so that we have a commutative diagram as

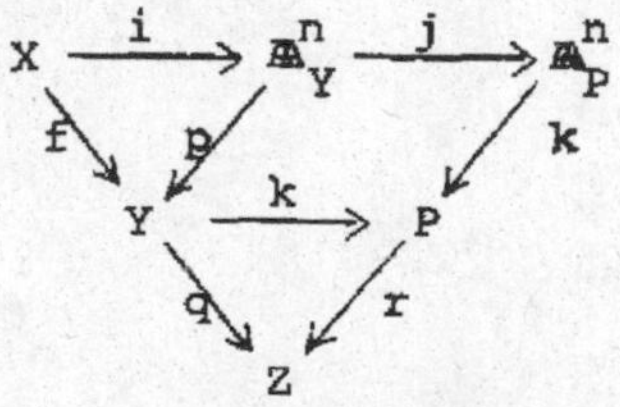

shown, with i,j,k closed immersions, p,q,r smooth, and the upper right square Cartesian. Define $c_{f,g}$ (depending on the charts chosen, and taking $ji: X \longrightarrow \mathbb{A}^n_P$ as a chart for gf) as follows:

$$c_{f,g}: (gf)^a \xrightarrow{\gamma_{ji,rq}} (ji)^y (rq)^z \xrightarrow{I,II} i^y j^y q^z r^z \xleftarrow{III} i^y p^z k^y r^z \xleftarrow{\gamma_{i,p}\,\gamma_{k,r}} f^b g^c .$$

Now we must show that $c_{f,g}$ is compatible with the permissible isomorphisms of change of chart for f and g. It is sufficient to vary one at a time. Furthermore, if $k': Y \longrightarrow Q$ is another chart for g, by considering the third chart $k'' = k \times k': Y \longrightarrow P \times_Z Q$, we reduce to the case where Q dominates P. So we have a diagram such as the one shown, and we must show that the two $c_{f,g}$ are compatible, i.e., that the following diagram is commutative:

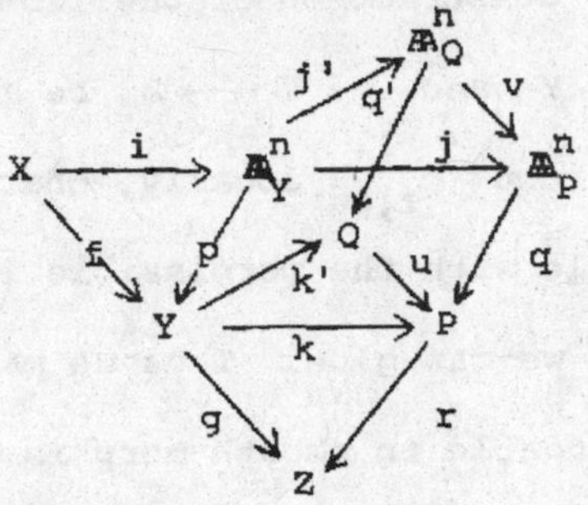

$$\begin{array}{ccccccccc}
(gf)^a & \xrightarrow{\gamma_{ji,rq}} & (ji)^Y(rq)^Z & \xrightarrow{\text{I, II}} & i^Y j^Y q^Z r^Z & \xleftarrow{\text{III}} & i^Y p^Z k^Y r^Z & \xleftarrow{\gamma_{k,r}} & i^Y p^Z g^c & \xleftarrow{\gamma_{i,p}} & f^b g^c \\
\downarrow{\scriptstyle \alpha_{gf}} & & & & & & & & & & \downarrow{\scriptstyle \alpha_g} \\
(gf)^d & \xrightarrow{\gamma_{j'i,vqr}} & (j'i)^Y(ruq')^Z & \xrightarrow{\text{I, II}} & i^Y j'^Y q'^Z (ru)^Z & \xleftarrow{\text{III}} & i^Y p^Z k'^Y (ru)^Z & \xleftarrow{\gamma_{k',ru}} & i^Y p^Z g^e & \xleftarrow{\gamma_{i,p}} & f^b g^e
\end{array}.$$

One may verify this (!) using the definition of a permissible isomorphism, and the compatibilities ii, v, and vii above.

Similarly we must consider a change of the chart on X to another, say $i'\colon X \longrightarrow \mathbb{A}^m_Y$, and as before we may assume that i' dominates i. This involves checking another analogous commutative diagram of isomorphisms, which we leave to the reader (!).

c) Construction of the isomorphism ψ_f for a finite morphism f. Let $f\colon X \longrightarrow Y$ be a finite morphism. As before, it will be sufficient to construct the isomorphism ψ_f locally on charts, provided our definition is compatible with permissible isomorphisms of charts. So let $i\colon X \longrightarrow P$ be a chart for f, and define

$$\begin{array}{ccc} X & \xrightarrow{i} & P \\ & {\scriptstyle f}\searrow \quad \swarrow{\scriptstyle p} & \\ & Y & \end{array}$$

$$\psi_f\colon \quad f^a \xrightarrow{\gamma_{i,p}} i^Y p^Z \xleftarrow{\text{IV}} f^Y .$$

If $j: X \longrightarrow Q$ is another chart, one may assume as usual that j dominates i, and one checks (!) using compatibility (v) above that ψ_f is compatible.

d) For a smooth morphism $g: X \longrightarrow Y$ we must construct the isomorphism φ_g. Take g as its own chart, and take $\varphi_g = \gamma$. There is no choice involved, hence nothing to check.

Having constructed the data a) - d), we must verify the conditions VAR 1 - VAR 5.

VAR 1). Given three morphisms f,g,h, we use a diagram such as the one shown to calculate $c_{f,g}$, etc., on the charts. The diagram of the condition becomes a large diagram whose commutative one checks (!) using (iii) and (iv) above.

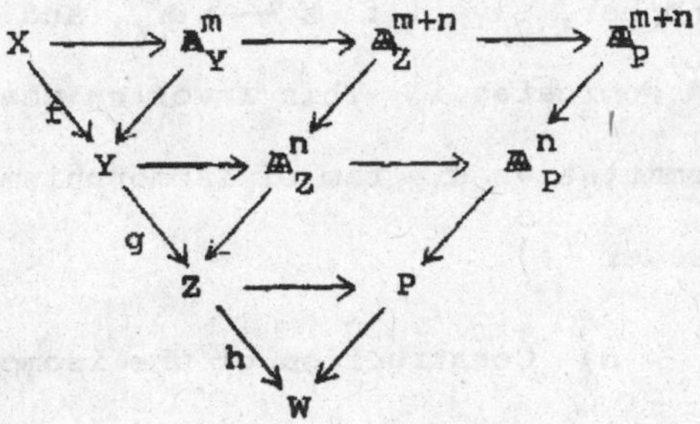

VAR 2). If f and g are consecutive finite morphisms, one checks (!) that $c_{f,g}$ is compatible with (I) via the isomorphisms ψ_f and ψ_g, by using (v) and (vi) above.

VAR 3). Trivial. One has only to observe that the isomorphism (III) is the identity if one side of the square is the identity.

VAR 4) and VAR 5) follow (!) from the definitions and a few more commutative diagrams.

This completes the proof of existence of the theory of variance, and we now show its uniqueness. For that purpose, suppose that $\{f^{\Delta}, c^{\Delta}_{f,g}, \psi^{\Delta}_{f}, \varphi^{\Delta}_{g}\}$ and $\{f^{\times}, c^{\times}_{f,g}, \psi^{\times}_{f}, \varphi^{\times}_{g}\}$ are two sets of data a)-d), each satisfying the conditions VAR 1 - VAR 5 (which we will call VAR 1^{Δ}, VAR $1^{\times}$, etc., to be precise). We will construct an isomorphism $\delta: f^{\Delta} \longrightarrow f^{\times}$, compatible with the data b)-d) and b')-d'), and we will observe that δ is unique.

Let $f: X \longrightarrow Y$ be a morphism of finite type, and choose locally an embedding $i: X \longrightarrow P$ into a scheme smooth over Y. Define δ by

$$\begin{array}{ccc} X & \xrightarrow{\ i\ } & P \\ & {\scriptstyle f}\searrow \quad \swarrow{\scriptstyle p} & \\ & Y & \end{array}$$

$$\delta: \ f^{\Delta} \xrightarrow{c^{\Delta}_{i,p}} i^{\Delta} p^{\Delta} \xrightarrow{\psi^{\Delta}_{i}, \varphi^{\Delta}_{p}} i^{Y} p^{Z} \xleftarrow{\psi^{\times}_{i}, \varphi^{\times}_{p}} i^{\times} p^{\times} \xleftarrow{c^{\times}_{i,p}} f^{\times} .$$

(Note that in order to be compatible with the $c_{i,p}$, ψ_{i}, φ_{p}, we must choose δ this way, which proves the uniqueness of δ.)

To see that δ is independent of the embedding chosen, let $j: X \longrightarrow Q$ be another one. Replacing Q by $P \times Q$ as usual we reduce to the case Q dominates P, and have a diagram such as the one shown. We must check that the perimeter of the following diagram is commutative:

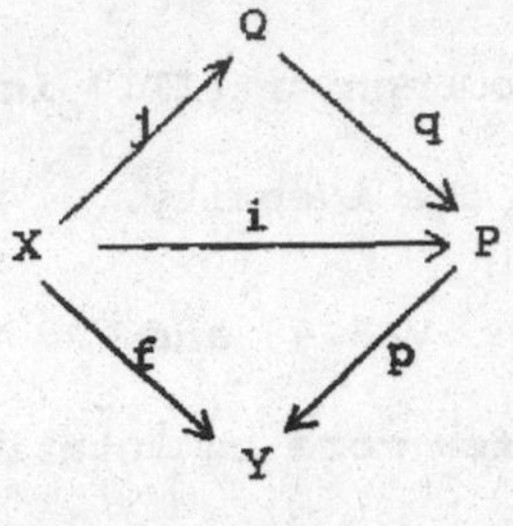

$$\begin{array}{ccccccccc}
f^{\Delta} & \xrightarrow{c^{\Delta}_{i,p}} & i^{\Delta}p^{\Delta} & \xrightarrow{\psi^{\Delta}_{i},\varphi^{\Delta}_{p}} & i^{Y}p^{Z} & \xleftarrow{\psi^{X}_{i},\varphi^{X}_{p}} & i^{X}p^{X} & \xleftarrow{c^{X}_{i,p}} & f^{X} \\
\Big\downarrow & \underline{\text{VAR } 1^{\Delta}} & \Big\downarrow c^{\Delta}_{j,q} & \underline{\text{VAR } 5^{\Delta}} & \Big\downarrow \text{IV} & & & & \Big\downarrow \\
\text{id} & & j^{\Delta}q^{\Delta}p^{\Delta} & \xrightarrow{\psi^{\Delta}_{j},\varphi^{\Delta}_{q},\varphi^{\Delta}_{p}} & j^{Y}q^{Z}p^{Z} & & \underline{\text{ditto}^{X}} & & \text{id} \\
\Big\downarrow & & \Big\uparrow c^{\Delta}_{p,q} & \underline{\text{VAR } 3^{\Delta}} & \Big\uparrow \text{II} & & & & \Big\downarrow \\
f^{\Delta} & \xrightarrow{c^{\Delta}_{j,pq}} & j^{\Delta}(pq)^{\Delta} & \xrightarrow{\psi^{\Delta}_{j},\varphi^{\Delta}_{pq}} & j^{Y}(pq)^{Z} & \xleftarrow{\psi^{X}_{j},\varphi^{X}_{pq}} & j^{X}(pq)^{X} & \xleftarrow{c^{X}_{j,pq}} & f^{X}
\end{array} .$$

That in fact it is, is shown by chopping it into little squares, which are commutative for the reasons shown. On the right we use VAR 1^{X}, VAR 3^{X}, and VAR 5^{X} similarly.

Thus δ is well-defined locally. Now cover X by open sets U_{ν} for which this is possible, and glue together the local isomorphisms

$$\delta_\nu: (f_{j_\nu})^\Delta \xrightarrow{\sim} (f_{j_\nu})^\times$$

where $j_\nu: U_\nu \longrightarrow X$ is the open immersion. Since j_ν^Δ is isomorphic via $\varphi_{j_\nu}^\Delta$ with j_ν^z which is the restriction, we can glue the δ_ν once we have checked (!) that the isomorphisms are compatible with restriction. Thus

$$\delta: f^\Delta \xrightarrow{\sim} f^\times$$

is defined.

Now we must check that δ is compatible with the isomorphisms $c_{f,g}$, ψ_f, φ_g. This can be done locally using the conditions VAR 1 - VAR 5, and we leave the details to the reader (!). q.e.d.

__Proposition 3.4.__ Let $f: X \longrightarrow Y$ be a morphism (with the conventions above) and let $K^\cdot$ be a residual complex on Y. Let d denote the codimension function on X (resp. Y) associated with the pointwise dualizing complex $f^\Delta K^\cdot$ (resp. $K^\cdot$). Then for each $x \in X$, if $y = f(x)$, we have

$$d(x) = d(y) + \text{tr.d. } k(x)/k(y).$$

Proof. The question is local, and compatible with composition of morphisms. Thus we reduce to the case f finite (trivial) or f smooth (which is [V 8.4]).

Corollary 3.5. Let $f: X \longrightarrow Y$ be a morphism (with the conventions above: f is of finite type, and the dimensions of its fibres are bounded). Then if Y admits a residual complex (resp. dualizing complex) so does X.

Proof. The existence of a residual complex on X, given one on Y, follows from the theorem. If Y admits a dualizing complex $R^{\cdot}$, then the associated codimension function d on Y is bounded. By the proposition, the codimension function associated to $Qf^{\Delta}E(R^{\cdot})$ is also bounded. But $Qf^{\Delta}E(R^{\cdot})$ is pointwise dualizing, hence dualizing (cf. proof of [V 8.2]).

§4. Trace for Residual Complexes.

In this section we define the trace map for residual complexes. For a morphism $f: X \longrightarrow Y$ (with the conventions of the previous section) it is a map of graded sheaves

$$\mathrm{Tr}_f: f_* f^{\Delta} K^{\cdot} \longrightarrow K^{\cdot}$$

where $K^{\cdot}$ is a residual complex on Y. It will be a morphism of complexes only if f is proper (see the Residue Theorem in the next chapter, which generalizes the classical theorem that the sum of the residues of a differential on a curve is zero).

First we define the trace map for a finite morphism, by carrying over the map of [III.6.5] to residual complexes. Let $f: X \longrightarrow Y$ be a finite morphism. We will denote by f' the functor

$$\bar{f}^* \underline{\mathrm{Hom}}_{\mathcal{O}_Y}(f_* \mathcal{O}_X, \cdot)$$

(using the notation of [III §6]) so that $f^{\flat} = \underline{\underline{R}} f'$.

Lemma 4.1. Let $f: X \longrightarrow Y$ be a finite morphism, and let $K^{\cdot}$ be a residual complex on Y. Then $f'(K^{\cdot})$ is a residual complex on X, and $f_* f'(K^{\cdot})$ is a Cousin complex on Y, with respect to the filtration $Z^{\cdot}$ associated with $Q(K^{\cdot})$ (cf. [IV §3]).

Proof. It is clear that f' takes quasi-coherent injectives on Y to quasi-coherent injectives on X, and it is also clear that $f'(K^{\cdot})$ has coherent cohomology, so we have only to check that it is isomorphic to a sum of sheaves of the form $J(x)$. Indeed, it will be sufficient to show, for each $y \in Y$, that

$$f'(J(y)) \cong \sum_{x \to y} J(x) .$$

Letting A be the local ring of y on Y, and B the semilocal ring of the points x of X lying over y, we have a local homomorphism $A \longrightarrow B$ where B is a finite A-module. If k is the residue field of A, and I is an injective hull of k over A, we must show that $J = \mathrm{Hom}_A(B,I)$ is a direct sum $\sum_{i=1}^{r} I_i$ where $k_1,\ldots,k_r$ are the residue fields of B, and I_i is an injective hull of k_i over B. For any B-module of finite type M, we have isomorphisms

$$\mathrm{Hom}_B(M,J) = \mathrm{Hom}_B(M, \mathrm{Hom}_A(B,I)) \cong \mathrm{Hom}_A(M,I) .$$

Now J is injective, and has support at the closed points of Spec B. Hence it is a direct sum of some number of copies of the injectives I_i. To find out how many, we have only to calculate

$$\mathrm{Hom}_B(k_i, J) \cong \mathrm{Hom}_A(k_i, I) \cong k_i ,$$

which is true since k_i is an A-module of finite length, and I is dualizing. Hence each I_i occurs just once.

Now $f_* f'(K^\cdot)$ is a direct sum, for each $y \in Y$, of

$$\sum_{x \rightarrow y} f_* J(x)$$

which is a constant sheaf spread out on $\{y\}^-$, and which occurs in degree $p = d(y)$, where d is the associated codimension function. Hence $f_* f'(K^\cdot)$ is a Cousin complex with respect to the filtration $Z^\cdot$ associated with $Q(K^\cdot)$.

q.e.d.

Now we are in a position to define the trace map for the finite morphism f, which we will call ρ_f to avoid confusion:

$$\rho_f : \ f_* f^Y(K^\cdot) \longrightarrow K^\cdot .$$

Let $K^{\cdot}$ be a residual complex on Y. Since it is an injective complex, the natural map

$$\xi_{f'}: \quad Qf'(K^{\cdot}) \longrightarrow f^{\flat}Q(K^{\cdot})$$

of the functor f' into its derived functor $f^{\flat}$ is an isomorphism in $D^{+}(X)$ [I.5.1]. Hence also

$$E\xi_{f'}: \quad f'(K^{\cdot}) = EQf'(K^{\cdot}) \longrightarrow Ef^{\flat}Q(K^{\cdot}) = f^{\Delta}(K^{\cdot})$$

is an isomorphism of residual complexes (recall $EQ \cong 1$ for any Cousin complex).

Consider the map

$$Qf_{*}f'(K^{\cdot}) \xrightarrow{\xi_{f_*}} \underline{\underline{R}}f_{*}Qf'(K^{\cdot}) \xrightarrow{\xi_{f'}} \underline{\underline{R}}f_{*}f^{\flat}Q(K^{\cdot}) \xrightarrow{\mathrm{Trf}_f} Q(K^{\cdot}),$$

where the last map, Trf_f, is the trace of [III 6.5]. By the lemma above, $f_{*}f'(K^{\cdot})$ is a Cousin complex with respect to the filtration $Z^{\cdot}$, and $K^{\cdot}$ is an injective Cousin complex, so by [III 3.2], this map in the derived category is represented by a unique map of complexes

$$f_{*}f'(K^{\cdot}) \longrightarrow K^{\cdot} \quad .$$

Composing with the inverse of the isomorphism $E\xi_{f'}$ above, we obtain the desired map

$$\rho_f\colon\ f_*f^Y(K^\cdot) \longrightarrow K^\cdot .$$

If $f\colon X \longrightarrow Y$, and $g\colon Y \longrightarrow Z$ are two finite morphisms, then there is a commutative diagram

$$\begin{array}{ccc} (gf)_*(gf)^Y & \xrightarrow{\ \rho_{gf}\ } & 1 \\ \Big\| I & & \Big\| \rho_g \\ g_*f_*f^Yg^Y & \xrightarrow{\ \rho_f\ } & g_*g^Y \end{array} \qquad \text{(viii)}$$

which follows from [III 6.6] and the functoriality of the above construction.

<u>Remark</u>. We have given the above definition in some detail to establish its relation with the trace map defined for the derived category in Chapter III. One could of course define ρ_f much more quickly by the usual "evaluation at one" map, without passing to the derived category, but we need the functorial properties below.

<u>Theorem 4.2</u>. As above, we work in the category of locally noetherian preschemes, and morphisms of finite type

such that the dimension of the fibres is bounded. There exists a unique theory of trace, consisting of the data a) below, subject to the conditions TRA 1 and TRA 2.

a) For each morphism $f: X \longrightarrow Y$, a morphism

$$Tr_f:\ f_* f^{\Delta} \longrightarrow 1$$

of functors from Res(Y) to the category of graded sheaves of $\mathcal{O}_Y$-modules (where 1 denotes the forgetful functor: consider a residual complex $K_Y^{\cdot}$ simply as a graded $\mathcal{O}_Y$-module).

TRA 1). If f and g are two consecutive morphisms of finite type, then there is a commutative diagram

$$\begin{array}{ccc} (gf)_*(gf)^{\Delta} & \xrightarrow{Tr_{gf}} & 1 \\ \Big\downarrow c_{f,g} & & \Big\downarrow Tr_g \\ g_* f_* f^{\Delta} g^{\Delta} & \longrightarrow & g_* g^{\Delta} \end{array} .$$

TRA 2). If f is a finite morphism, then Tr_f is the one we already know, i.e., $Tr_f = \rho_f \psi_f$, using the notation of Theorem 3.1.

Lemma 4.3. [EGA IV § ?] Let $f: X \longrightarrow Y$ be a morphism of finite type, let $x \in X$ be a point which is closed in its fibre, and let Z be the closure of x in X, with the reduced induced structure. Then there is an open neighborhood V of $y = f(x)$ such that $Z \cap f^{-1}(V)$ is finite over V.

Proof. Replacing X by Z and Y by $f(Z)^-$ with the reduced induced strucutre, we reduce to the case where X and Y are integral schemes, x is the generic point of X, and f is dominant. Since furthermore the question is local on X and Y, we may assume X and Y are affine, and thus we reduce to the following statement in algebra:

Let $A \longrightarrow B$ be an inclusion of integral domains, with B a finitely generated A-algebra. Let K be the quotient field of A, and assume that $B \otimes_A K$ is a finite extension field of K. Then there exists an element $f \neq 0$ in A such that $B \otimes_A A_f$ is a finite A_f-module.

To prove this statement, let $b_1, \cdots, b_n$ be a finite set of elements in B such that

1) the b_i generate B as an A-algebra, and

2) the elements $b_i \otimes 1$ generate $B \otimes K$ as a K-vector space.

Then for each i,j, $b_i b_j \otimes 1 \in B \otimes K$, so we can write

$$b_i b_j \otimes 1 = \sum_{k=1}^{n} b_k \otimes \lambda_{ijk}, \qquad \lambda_{ijk} \in K.$$

Let f be a common denominator for all the λ_{ijk}, $1 \leq i,j,k \leq n$. Then the λ_{ijk} are all in A_f, so the A_f-module B_o generated by the $b_i \otimes 1$ is in fact a ring. But B_o contains a set of generators for $B \otimes A_f$ as an A_f-algebra, so $B_o = B \otimes A_f$, which shows that the latter is a finitely generated A_f-module. q.e.d.

<u>Lemma 4.4</u>. Let $f\colon X \longrightarrow Y$ be a morphism, and let $K^{\cdot}$ be a residual complex on Y. Then there is a unique map

$$Tr_f\colon \; f_* f^{\Delta} K^{\cdot} \longrightarrow K^{\cdot}$$

of graded sheaves on Y, such that whenever U is an open subset of Y, and W is a closed subscheme of $f^{-1}(U)$, finite over U, we have a commutative diagram

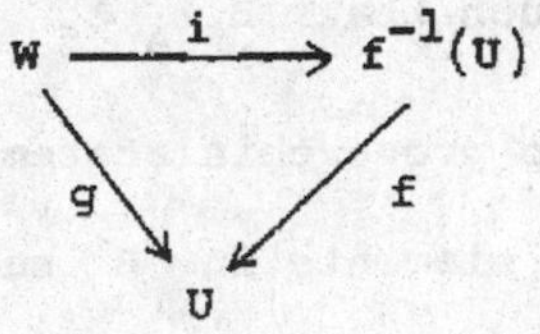

$$g_*g^Y(K^\cdot) \xleftarrow{\psi_g} g_*g^\Delta(K^\cdot) \xrightarrow{c_{i,f}} f_*i_*i^\Delta f^\Delta(K^\cdot) \xrightarrow{\psi_i} f_*i_*i^Y f^\Delta(K^\cdot) \xrightarrow{\rho_i} f_*f^\Delta(K^\cdot)$$

$$g_*g^Y(K^\cdot) \xrightarrow{\rho_g} K^\cdot_Y \xleftarrow{Tr_f} f_*f^\Delta(K^\cdot)$$

(Here we denote by f the restriction of f to $f^{-1}(U)$, by $K^\cdot$ the restriction of $K^\cdot$ to V, and so forth.)

<u>Proof</u>. Let d denote the codimension function on Y associated with the residual complex $K^\cdot$, and let it also denote the codimension function on X associated with the residual complex $f^\Delta K^\cdot$ (the one we mean will always be clear from the context). Then according to the definition of the residual complexes we have isomorphisms

$$K^p \cong \sum_{d(y)=p} J(y)$$

and

$$(f^\Delta K^\cdot)^p \cong \sum_{d(x)=p} J(x) .$$

We will denote by $J(y)$ (resp. $J(x)$) the unique subsheaf of K^p

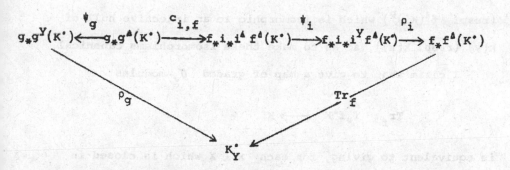

(resp. $f^{\Delta}(K^{\cdot})^{p}$) which is isomorphic to an injective hull of $k(y)$ (resp. $k(x)$), so as to make these isomorphisms canonical.

I claim that to give a map of graded $\mathcal{O}_Y$-modules

$$\mathrm{Tr}_f\colon\ f_*f^{\Delta}K^{\cdot} \longrightarrow K^{\cdot}$$

is equivalent to giving, for each $x \in X$ which is closed in its fibre, a map

$$\mathrm{Tr}_{f,x}\colon\ f_*J(x) \longrightarrow J(y)$$

where $y = f(x)$. Indeed, x is closed in its fibre if and only if $\mathrm{tr.d.}\ k(x)/k(y) = 0$, i.e., if and only if $d(x) = d(y)$, by Proposition 3.4 above. Hence Tr_f certainly gives rise to a collection of maps $\mathrm{Tr}_{f,x}$ as above. On the other hand, if $y' \in Y$ is a point not equal to y, with $d(y') = d(x)$, then there is no non-zero map of $f_*J(x)$ into $J(y')$, because $d(y') \geq d(y)$ by Proposition 3.4, and so $y' \notin \{y\}^-$. (Note that $f_*J(x)$ has support in $\{y\}^-$). Thus Tr_f is determined by the maps $\mathrm{Tr}_{f,x}$ above, and these can be given arbitrarily.

We will now construct maps $\mathrm{Tr}_{f,x}$ as above for each point $x \in X$ which is closed in its fibre.

Given an $x \in X$ closed in its fibre, choose by Lemma 4.3 an open neighborhood V of $y = f(x)$ such that $Z = \{x\}^{-} \cap f^{-1}(V)$ is finite over V. Let I be the ideal of Z with the reduced induced structure, and let Z_n, $n = 1,2,\cdots$, be the subscheme of X defined by I^n. For $n < n'$ we have a closed immersion $j_{nn'}: Z_n \longrightarrow Z_{n'}$, and one can verify (!) using VAR 1, VAR 2, and (viii) above, that the following diagram is commutative:

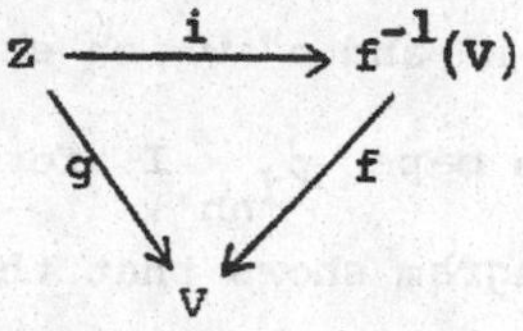

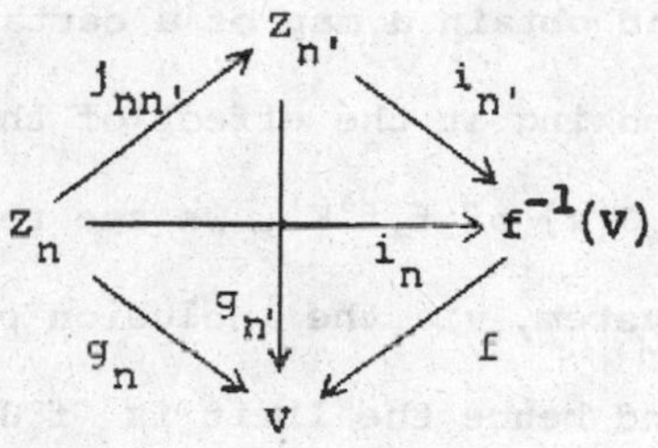

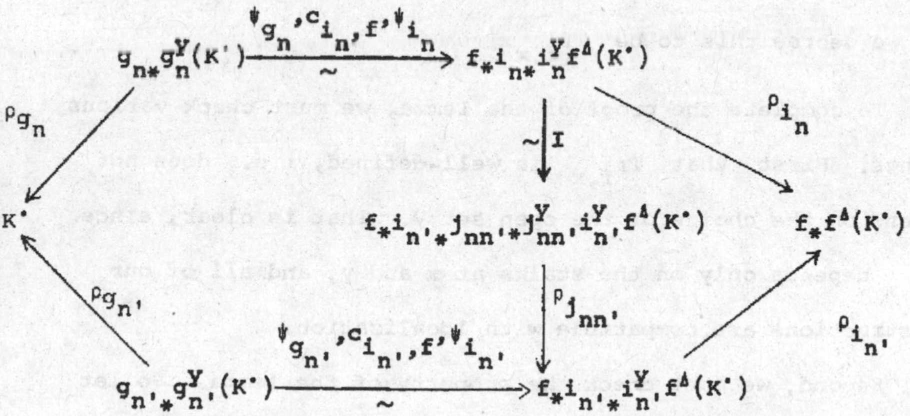

.

One can also write down another diagram with three indices n, n' and n'' which shows that the sheaves $f_* i_{n*} i_n^Y f^\Delta(K^\cdot)$ and the maps $\rho_{j_{nn'}}$, I form a direct system as n varies, and this diagram shows that the ρ_{i_n} map this direct system into $f_* f^\Delta(K_Y^\cdot)$ and the $\rho_{g_n} \psi_{g_n} c_{i_n}, f^\psi i_n$ map it to $K^\cdot$. Furthermore the ρ_{i_n} are all injective maps. Thus we can pass to the limit and obtain a map of a certain subsheaf of $f_* f^\Delta(K_Y^\cdot)$ to $K_Y^\cdot$. Looking at the effect of this construction on the component $f_* J(x)$ of $f_* f^\Delta K^\cdot$, we see that the nth term of the direct system, via the inclusion ρ_{i_n}, is just $f_* \underline{Hom}_{\mathcal{O}_X}(\mathcal{O}_X/I^n, J(x))$, and hence the limit is $f_* J(x)$ itself, since $J(x)$ has support on Z. Hence the map we obtain is defined on all of $f_* J(x)$, and we decree this to be $Tr_{f,x}$.

To complete the proof of the lemma, we must check various things. First, that $Tr_{f,x}$ is well-defined, i.e., does not depend on the choice of the open set V. That is clear, since $Tr_{f,x}$ depends only on the stalks at x and y, and all of our constructions are compatible with localization.

Second, we must check the property of the lemma. So let W and U be as in the statement. It will be sufficient to check the diagram on the component of x for each $x \in W$. Let Z be the closure of x, with the reduced induced structure (which will be finite over U). Let J be the ideal of Z in W, and for each $n = 1,2,\ldots$, let Z_n' be the subscheme of W defined by J^n. Let Z_n be as above. Then there are closed immersions as shown. We leave to the reader (!) to write down a huge commutative diagram of Z_n's and Z_n''s which in the limit gives the diagram of the lemma on the component of x.

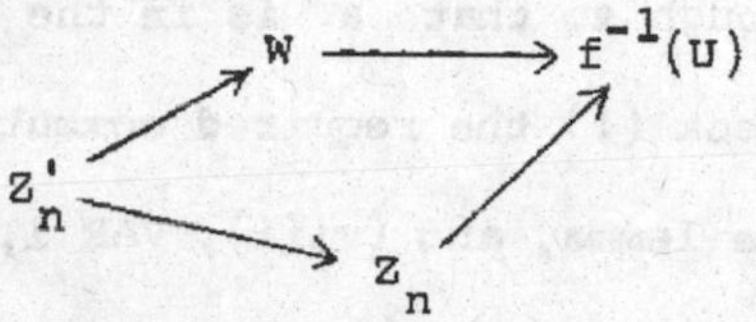

Third we must check the uniqueness, but this is clear from the construction.

<u>Proof of theorem</u>. For a morphism $f: X \longrightarrow Y$, we define Tr_f to be the map given by the lemma. It is clearly functorial in $K^\cdot$. In case f is a finite morphism, we can take U = Y and W = X, so that $Tr_f = \rho_f \psi_f$, which proves condition TRA 2.

To prove TRA 1, let $f: X \longrightarrow Y$ and $g: Y \longrightarrow Z$ be two morphisms. It is sufficient to prove the condition of TRA 1 for a single residual complex $K^\cdot$ on X and for a single $x \in X$ which is closed in its fibre over Z. The question is local, so we may assume that $W = \{x\}^-$ is finite over Z, and that $V = \{y\}^-$,

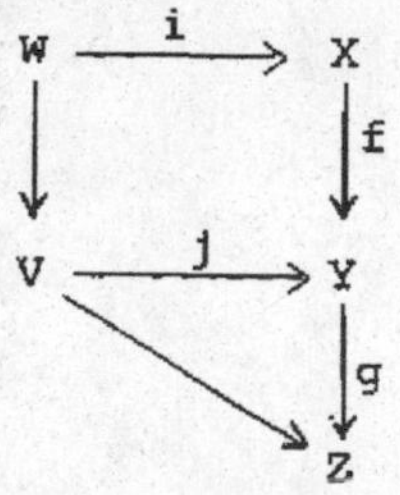

where $y = f(x)$, is finite over Z. Finally it is sufficient to check the commutativity for a given element $a \in \Gamma(J(x))$. We choose a subscheme structure W_n on W with n large enough so that a is in the image of $i^{Y}J(x)$. Now one can check (!) the required commutativity using the property of the lemma, and (viii), VAR 1, and VAR 2 above.

The uniqueness of Tr_f is clear, as it was in the lemma, since TRA 1 and TRA 2 imply the condition of the lemma.

q.e.d.

§5. Behavior with respect to certain base changes.

In this section we show that the functor f^Δ and the morphism Tr_f are compatible with certain base changes which take residual complexes into residual complexes.

Definition. A morphism $f: X \longrightarrow Y$ of locally noetherian preschemes is residually stable if

a) f is flat

b) f is integral [EGA II 6.1.1], and

c) the fibres of f are Gorenstein preschemes

[V §9].

Examples. 1. An open immersion is residually stable.

2. A composition of residually stable morphisms is residually stable (use [V 9.6]).

3. If X and Y are the spectra of fields $k \subseteq K$, then f is residually stable if and only if K is algebraic over k.

4. If $f: X \longrightarrow Y$ is residually stable, then every point $x \in X$ is closed in its fibre, since f is integral. Thus the fibres are zero-dimensional Gorenstein preschemes.

Proposition 5.1. If $f: X \longrightarrow Y$ is a residually stable morphism, and if $u: Y' \longrightarrow Y$ is a morphism of finite type, then $g: X' = X \times_Y Y' \longrightarrow Y'$ is also residually stable.

Proof. Clearly g is flat and integral. To show that the fibres of g are Gorenstein, let $y' \in Y$, let $y = u(y')$, and consider the map of fibres $v: X'_{y'} \longrightarrow X_y$. Now X_y is Gorenstein since f is residually stable. The fibres of v are tensor products of fields, one of which is finitely generated (namely $k(y')/k(y)$), hence Gorenstein [V 9.5]. Hence $X'_{y'}$ is Gorenstein [V 9.6].

Lemma 5.2. Let A be a noetherian local ring, and let I be an A-module. Then I is an injective hull of the residue field k of A if and only if

a) I has support at the closed point of Spec A,

b) $\mathrm{Hom}_A(k,I) \cong k$, and

c) There is a sequence of ideals $\mathfrak{a}_1 \supseteq \mathfrak{a}_2 \supseteq \dots$ of A which form a base for the $\mathfrak{m}$-adic topology, such that for all n,

$$\text{length}\,(A/\mathfrak{a}_n) = \text{length}\,(\mathrm{Hom}_A(A/\mathfrak{a}_n, I)).$$

Proof. If I is an injective hull of k, then a),b),c) are immediate, since $\mathrm{Hom}_A(\cdot,I)$ is a dualizing functor for modules of finite length. In fact, c) holds for any ideal which is primary for $\mathfrak{m}$.

Conversely, let I be an A-module satisfying a),b),c), and let J be an injective hull of k. By b), we can find an injection $k \subseteq I$, and since J is injective, the natural map $\theta: k \longrightarrow J$ extends to a map $\psi: I \longrightarrow J$. I claim ψ is injective. Indeed, given $y \in I$, $y \neq 0$, we have $\mathfrak{m}^n y = 0$ for sufficiently large n, by a). Let n be the least such integer. Then there is an $x \in \mathfrak{m}^{n-1}$ for which $xy \neq 0$. But $\mathfrak{m}(xy) = 0$, so $xy \in k$. Hence $\psi(xy) = \theta(xy) \neq 0$. But $\psi(xy) = x \cdot \psi(y)$, so $\psi(y) \neq 0$, and ψ is injective.

Now choose a sequence of ideals $\mathfrak{a}_1 \supseteq \mathfrak{a}_2 \supseteq \ldots$ as in c). Then we note that

$$\psi(\mathrm{Hom}(A/\mathfrak{a}_n, I)) \subseteq \mathrm{Hom}(A/\mathfrak{a}_n, J),$$

and both have the same length, namely the length of $A/\mathfrak{a}_n$. Therefore they are equal. But I and J are the union of the submodules of elements annihilated by $\mathfrak{a}_n$, so we see that ψ is also surjective. Thus I is isomorphic to J, and so is an injective hull of k.

Proposition 5.3. Let $f\colon X \longrightarrow Y$ be a residually stable morphism, and let $K^{\cdot}$ be a residual complex on Y. Then $f^{*}(K^{\cdot})$ is a residual complex on X.

Proof. Clearly $f^{*}(K^{\cdot})$ is a complex of quasi-coherent sheaves with coherent cohomology. We have only to check that there is an isomorphism

$$\sum_{p} f^{*}(K^{p}) \cong \sum_{x \in X} J(x).$$

Thus we reduce to the following local statement: For each $x \in X$, let $y = f(x)$, and let I be an injective hull of $k = k(y)$ over the local ring $A = \mathcal{O}_y$ of y. Let $B = \mathcal{O}_x$ be the local ring of x. Then $I \otimes_A B$ is an injective hull of $k' = k(x)$ over B.

We apply the criteria of the lemma above. Since I is an injective hull of k over A, we have

a) I has support at the closed point of Spec A

b) $\mathrm{Hom}_A(k,I) \cong k$

c) For each n,

$$\text{length } (A/\mathfrak{m}^{n}) = \text{length } (\mathrm{Hom}_A(A/\mathfrak{m}^{n}, I)),$$

where $\mathfrak{m}$ is the maximal ideal of A.

Now since x is closed in its fibre, $I \otimes_A B$ has support at the closed point of Spec B. Since B is flat over A, we have

$$\mathrm{Hom}_B(B/\mathfrak{m}B, I\otimes_A B) \cong B/\mathfrak{m}B.$$

Therefore

$$\mathrm{Hom}_B(k', I\otimes_A B) \cong \mathrm{Hom}_B(k', B/\mathfrak{m}B) \cong k' ,$$

since $B/\mathfrak{m}B$ is an Artinian Gorenstein ring. Finally note that for each n,

$$\text{length } (B/\mathfrak{m}^n B) = \text{length } (\mathrm{Hom}_B(B/\mathfrak{m}^n B, I\otimes_A B)).$$

Thus the conditions of the lemma are fulfilled, and $I\otimes_A B$ is an injective hull of k' over B.

Proposition 5.4. Let Y be the spectrum of a local Artin ring A. Then one can find a residually stable morphism $u\colon Y' \longrightarrow Y$ where Y' is the spectrum of a local Artin ring A' with algebraically closed residue field.

Proof. By [EGA 0_{III} 10.3.1] one can find a local Artin ring A' and a local homomorphism $A \longrightarrow A'$ such that A' is flat over A, $\mathfrak{m}A' = \mathfrak{m}'$, and k' = the algebraic closure of k. Then Y' = Spec A' will do. Indeed, Y' is flat over Y. It is integral since it is Artinian, and its residue field is algebraic over that of Y. The only fibre is a field, which is Gorenstein.

Now we come to the behavior of f^{Δ} and Tr_f under residually stable base change. Let $f: X \longrightarrow Y$ be a morphism of finite type, and let $u: Y' \longrightarrow Y$ be a residually stable morphism. Let $X' = X \times_Y Y'$, and let u', f' be as shown.

$$\begin{array}{ccc} X' & \xrightarrow{u'} & X \\ \downarrow{\scriptstyle f'} & & \downarrow{\scriptstyle f} \\ Y' & \xrightarrow{u} & Y \end{array}$$

If f is a finite morphism, there is a natural isomorphism

$$f'^{Y} u^* \xrightarrow{\sim} u'^* f^{Y} \qquad \text{(V)}$$

derived from [III 6.3] (cf. section 2 above).

On the other hand if f is smooth, there is a natural isomorphism

$$f'^{Z} u^* \xrightarrow{\sim} u'^* f^{Z} \qquad \text{(VI)}$$

derived from [III 2.1].

<u>Theorem 5.5</u>. For every morphism $f: X \longrightarrow Y$ of finite type, and every residually stable morphism $u: Y' \longrightarrow Y$ (using the notations above) there is an isomorphism

$$d_{u,f}: \; f'^{\Delta} u^* \xrightarrow{\sim} u'^* f^{\Delta}$$

such that

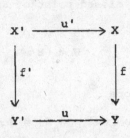

1) If $v: Y'' \longrightarrow Y'$ is another residually stable morphism, then $d_{uv,f} = d_{u,f} d_{v,f'}$.

2) If f,g are two consecutive morphisms of finite type, then the isomorphisms d are compatible with $c_{f,g}$ and $c_{f',g'}$.

3) If f is a finite morphism, then $d_{u,f}$ is compatible with V via the isomorphisms $\psi_f, \psi_{f'}$.

4) If f is a smooth morphism, then $d_{u,f}$ is compatible with VI via the isomorphisms $\varphi_f, \varphi_{f'}$.

<u>Suggestion of Proof (!).</u> Show first that V and VI are compatible with the isomorphisms I, II, III and IV of §2. Define $d_{u,f}$ for f finite or smooth using 3) and 4). Then define $d_{u,f}$ for arbitrary f and check its properties, following the construction of f^{Δ} given in §3.

<u>Theorem 5.6.</u> Let u and f be as in the previous theorem. Then there is a commutative diagram

$$\begin{array}{ccc} f'_* f'^{\Delta} u^* & \xrightarrow{\ Tr_{f'}\ } & u^* \\ \downarrow{\scriptstyle d_{u,f}} & & \uparrow{\scriptstyle Tr_f} \\ f'_* u'^* f^{\Delta} & = & u^* f_* f^{\Delta} \end{array} .$$

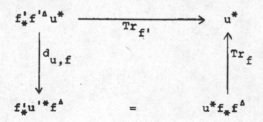

<u>Proof (!)</u>. Show first that the analogous diagram of f^{v}, ρ_f and V is valid for f finite. Then follow the construction of Tr_f in §4 to show that it is true in general.

CHAPTER VII. THE DUALITY THEOREM

§1. Curves over an Artin ring.

In this section we will make explicit the residual complex on a curve over the spectrum of an Artin ring, and we will identify the trace map of [VI §4] with the "classical" residue of a differential. Then, in the case of the projective line over an Artin ring with algebraically closed residue field, we will prove that the sum of the residues is zero, i.e., the trace map is a morphism of complexes. This special case will be used in the following section to prove the general residue theorem for a proper morphism of locally noetherian preschemes, which in turn implies that the sum of the residues is zero on any proper curve.

Throughout this section we will let Y be the spectrum of a local Artin ring A with residue field k, and we let X be a smooth curve over Y (i.e., a connected irreducible prescheme smooth over Y, with relative dimension one). A closed point $x \in X$ is rational over Y if its residue field $k(x)$ is k. In that case one can find a local parameter $t \in \mathcal{O}_x$ with the following properties:

(0) $\mathcal{O}_x/t \cong A$.

(1) $t \in \mathfrak{m}_x - \mathfrak{m}_x^2$

(2) t is a non-zero-divisor in $\mathcal{O}_x$

(3) $\mathcal{O}_x/t^n$ is a free A-module with basis $1,t,\ldots,t^{n-1}$

(4) The total quotient ring K of $\mathcal{O}_x$ is generated as an $\mathcal{O}_x$-module by $1,t^{-1},t^{-2},\ldots$

Proposition 1.1. Let $f: X \longrightarrow Y$ be as above. Let I be an injective hull of k over A, so that $\tilde{I}$ is a residual complex on Y. Then

a). $f^{z}(\tilde{I})$ is the complex

$$\overset{-1}{i_\eta(I \otimes_A \omega_\eta)} \longrightarrow \overset{0}{\coprod_{\substack{x \in X \\ \text{closed}}} i_x(H_x^1(I \otimes_A \omega_x))}$$

where $\omega = \Omega^1_{X/Y}$ is the sheaf of relative 1-differentials, η is the generic point of X, i_η, i_x is the notation of [II §7], meaning the given module spread out as a constant sheaf, and H_x^1 is a local cohomology group [IV §1].

b). $f^{z}(\tilde{I})$ is an injective resolution of $f^*(\tilde{I}) \otimes \omega[1]$.

c). If $x \in X$ is a point rational over Y, then

$$H^1_x(I \otimes_A \omega_x) \cong I \otimes_A \omega_x \otimes K/\mathcal{O}_x$$

where K is the total quotient ring of $\mathcal{O}_x$ (which is equal to the stalk $\mathcal{O}_\eta$ at the generic point, hence independent of x).

Proof. a). follows directly from the definition of f^z [VI §2], using [IV §3] and [IV.1.F].

b). follows from [V 8.3],[V.7.3], and [IV.3.1].

c). Suppose that x is rational over Y, and let $t \in \mathcal{O}_x$ be a local parameter. Then by [V.4.1] we can calculate the local cohomology as

$$H^1_x(I \otimes_A \omega_x) = \varinjlim_n \mathrm{Ext}^1_{\mathcal{O}_x}(\mathcal{O}_x/t^n, I\otimes_A\omega_x) .$$

But t is a non-zero-divisor in $\mathcal{O}_x$, so we can calculate the Ext with the resolution

$$0 \longrightarrow \mathcal{O}_x \xrightarrow{t^n} \mathcal{O}_x \longrightarrow \mathcal{O}_x/t^n \longrightarrow 0 ,$$

and find that

$$\mathrm{Ext}^1_{\mathcal{O}_x}(\mathcal{O}_x/t^n, I\otimes_A\omega_x) \cong I \otimes_A \omega_x/t^n(I\otimes_A\omega_x) .$$

The map in the direct system is multiplication by t, and

$$\varinjlim_{n} \mathcal{O}_x/t^n \cong K/\mathcal{O}_x ,$$

so we have

$$H^1_x(I \otimes_A \omega_x) \cong I \otimes_A \omega_x \otimes K/\mathcal{O}_x .$$

*   *   *

Now let Z be a closed subscheme of X, concentrated at a closed point $x_o \in X$, rational over Y. Then we can write $Z = \operatorname{Spec} B$, with B a local Artin ring, finite over A. Let I be an injective hull of k over A as above, and take $\tilde{I}$ to be a residual complex on Y.

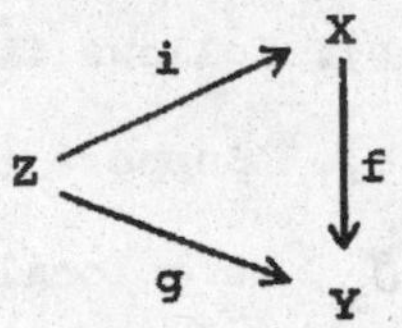

We propose to calculate explicitly the residue isomorphism (IV) of [VI §2] between $i^Y f^Z(\tilde{I})$ and $g^Y(\tilde{I})$. Since Z is an affine scheme, we will use modules, instead of sheaves, for convenience. Using the results above, we have

$$\Gamma(g^Y(\tilde{I})) = \operatorname{Hom}_A(B, I)$$

$$\Gamma(i^Y f^Z(\tilde{I})) = \operatorname{Hom}_{\mathcal{O}}(B, I \otimes_A \omega_{x_o} \otimes K/\mathcal{O})$$

where $\mathcal{O} = \mathcal{O}_{x_o}$. Indeed, since Z is concentrated at the point x_o, there are no homomorphisms of $\mathcal{O}_Z$ into $i_\eta(I \otimes_A \omega_\eta)$ or into $i_x(H^1_x(I \otimes_A \omega_x))$ for $x \neq x_o$.

Let t be a local parameter at x_o. Let $b_o \in \mathfrak{m}_B$ be the image of t under the structural morphism $\mathcal{O} \longrightarrow B$. Finally, note that ω_{x_o} is a free $\mathcal{O}$-module with basis dt, so we can represent elements of $I \otimes_A \omega_{x_o} \otimes K/\mathcal{O}$ as

$$\sum_{i<0} a_i t^i dt$$

with $a_i \in I$. Using this notation, we have the following

<u>Lemma 1.2</u>. If

$$\varphi \in \mathrm{Hom}_A (B, I)$$

and $$\psi \in \mathrm{Hom}_{\mathcal{O}} (B, I \otimes \omega_x \otimes K/\mathcal{O})$$

correspond under the residue isomorphism, then for each $b \in B$,

$$\psi(b) = \sum_{r=0}^{\infty} \varphi(bb_o^r) t^{-r-1} dt$$

and $$\varphi(b) = \text{coefficient of } t^{-1}dt \text{ in } \psi(b).$$

(Note that the sum in the first expression is finite since b_o is nilpotent.)

Proof. Left as a good exercise in definition-chasing to the reader!

Proposition 1.3. With the above notations, the trace map of [VI §4]

$$\mathrm{Tr}_f:\ f_* f^{\Delta}(\tilde{I}) \longrightarrow \tilde{I}$$

is zero on $f_* i_\eta(I \otimes_A \omega_\eta)$ (where we have identified $f^{\Delta}(\tilde{I})$ with $f^Z(\tilde{I})$) and for each closed point $x \in X$, rational over Y, its restriction

$$\Gamma(\mathrm{Tr}_{f,x}):\ I \otimes_A \omega_x \otimes K/\mathcal{O}_x \longrightarrow I$$

to the stalk at x is given as follows: let $t \in \mathcal{O}_x$ be a local parameter, and write $u \in I \otimes_A \omega_x \otimes K/\mathcal{O}_x$ as

$$u = \sum_{i<0} a_i t^i dt$$

with $a_i \in I$. Then $\Gamma(\mathrm{Tr}_{f,x})(u) = a_{-1}$.

Proof. To apply the definition of the trace map, we must choose a closed subscheme $i: Z \longrightarrow X$ of X, finite over Y, such that $u \in \Gamma(i^Y f^Z(\tilde{I}))$. Given u as above, suppose that $a_i = 0$ for $i < -r$. Then $Z = \mathrm{Spec}\ \mathcal{O}_x/t^r$ will do. Our element u is then identified with the element

$$\psi \in \mathrm{Hom}_{\mathcal{O}_x}(B, I \otimes_A \omega_x \otimes K/\mathcal{O}_x)$$

defined by $\psi(1) = u$. This corresponds by the residue isomorphism to a certain

$$\varphi \in \mathrm{Hom}_A(B, I) ,$$

and by definition of the trace morphism we then have

$$\Gamma(\mathrm{Tr}_{f,x})(u) = \varphi(1) .$$

But by the lemma above, $\varphi(1)$ is the coefficient of $t^{-1}dt$ in $\psi(1) = u$, so our trace is a_{-1}, as required.

Remark. Since the trace morphism has already been defined, and we are here calculating it, we can state as a corollary that the process of taking the coefficient of $t^{-1}dt$ in u is independent of the local parameter t chosen. This is in strong contrast with the approach of Serre's book [16, Ch. II §7], where the residue of a differential is *defined* as a_{-1}, and a tedious proof is required to show that it is independent of the local parameter [loc. cit. Prop. 5]. To establish the relation between his approach and ours, we make the following

<u>Definition</u>. Let M be an A-module, and let

$$w \in M \otimes_A \omega_\eta$$

be a meromorphic differential on X with coefficients in M. Then for each closed point $x \in X$, rational over Y, we define the <u>residue</u> of w at x, as follows. Choose a local parameter t at x. Let

$$w_x \in M \otimes_A \omega_x \otimes K/\mathcal{O}_x$$

be the image of w under the natural map

$$\omega_\eta = \omega_x \otimes K \longrightarrow \omega_x \otimes K/\mathcal{O}_x .$$

Write

$$w_x = \sum_{i<0} a_i t^i dt$$

with $a_i \in M$, and then define

$$\mathrm{Res}_x(w) = a_{-1} .$$

(Observe that a_{-1} is independent of the choice of local parameter. Indeed, we may assume that M is finitely generated, in which case it is a submodule of I^n for some n, so we reduce to the case $M = I$, where $a_{-1} = \Gamma(Tr_{f,x})(w_x)$ which is independent of t.)

Lemma 1.4. Residues have the following properties [16, Ch. II §7].

(i) $Res_x(w)$ is A-linear in w.

(ii) $Res_x(w) = 0$ if $w \in M \otimes \omega_x$.

(iii) $Res_x(dg) = 0$ for any $g \in M \otimes_A K$.

(iv) $Res_x(t^{-1}dt) = 1$ if t is a local parameter at x.

Proof. All properties follow from the definition.

Theorem 1.5 (Residue formula for the projective line). Let A be a local Artin ring with algebraically closed residue field k, let Y = Spec A, and let $X = \mathbb{P}^1_Y$ be the projective line over Y. Let M be an A-module, and let

$$w \in M \otimes_A \omega_\eta$$

be a meromorphic differential on X with coefficients in M. Then

$$\sum_{\substack{x \in X \\ \text{closed}}} Res_x(w) = 0 .$$

Proof. (Modeled on the proof of [16, Ch. II, §12, Lemma 3], which is the same statement in the case A = k = M.) First of all, we may assume that M is finitely generated. Then writing M as a quotient of A^n for suitable n, we reduce to the case M = A.

Let $w \in \omega_\eta$ be a meromorphic differential (recall η is the generic point of X). Let X_o be the affine line Spec A[t] contained as an open subset of X. Then from the exact sequence of sheaves

$$0 \longrightarrow \omega \longrightarrow i_\eta(\omega_\eta) \longrightarrow \coprod_{\substack{x \in X \\ \text{closed}}} i_x(\omega_x \otimes K/\mathcal{O}_x) \longrightarrow 0$$

we obtain the following exact sequence of global sections on X_o:

$$0 \longrightarrow \Gamma(X_o, \omega) \longrightarrow \omega_\eta \longrightarrow \coprod_{\substack{x \in X_o \\ \text{closed}}} \omega_x \otimes K/\mathcal{O}_x \longrightarrow 0 .$$

For each closed point $x \in X_o$, we can choose a local parameter of the form $t-c_x$, with $c_x \in A$. Then our exact sequence shows that w can be written (uniquely) in the form

$$w = f(t)dt + \sum_{\substack{x \in X_o \\ \text{closed}}} \sum_{i<0} a_{i,x}(t-c_x)^i dt$$

where $f(t) \in A[t]$ is a polynomial in t with coefficients in A. By linearity, we reduce to proving the residue formula in the two cases

(1) $w = t^n dt$ $\quad n \geq 0$

(2) $w = (t-c_{x_o})^n dt$ $\quad n < 0, \ x_o \in X_o$.

In the first case, $\mathrm{Res}_x(w) = 0$ for every $x \in X_o$, by (ii) of the lemma. We take $s = 1/t$ as a local parameter at ∞ (the unique point of $X-X_o$). Then $w = -s^{-n-2}ds$, and $n-2 < -1$, so the residue is zero there also.

In the second case, for any $x \neq x_o$, $x \in X_o$, $t-c_{x_o}$ is invertible in $\mathcal{O}_{x_o}$, so the residue at x is zero. The residue at x_o is 0 if $n < -1$, 1 if $n = -1$. We take $s = 1/(t-c_{x_o})$ as a local parameter at ∞. Then $w = -s^{-n-2}ds$, so its residue is 0 if $n < -1$, and -1 if $n = -1$. Hence the sum of the residues is zero. q.e.d.

<u>Remark</u>. It follows from the theorem of the next section that the theorem remains true for any smooth curve X <u>proper</u> over Y.

<u>Corollary 1.6</u>. Under the hypotheses of the theorem

$$\mathrm{Tr}_f\colon\ f_*f^\Delta(\check{Y}) \longrightarrow \check{Y}$$

is a morphism of complexes.

<u>Proof</u>. Using Proposition 1.3 above, this is precisely the statement of the Theorem in the case M = I.

§2. The residue theorem.

Theorem 2.1. (Residue theorem) Let $f\colon X \longrightarrow Y$ be a proper morphism of noetherian schemes, and let $K^{\cdot}$ be a residual complex on Y. Then the trace map

$$\mathrm{Tr}_f\colon\ f_* f^{\Delta} K^{\cdot} \longrightarrow K^{\cdot}$$

defined in [VI §4] is a morphism of complexes.

Proof. We write the residual complexes as sums of injective hulls,

$$\sum_p K^p = \coprod_{y \in Y} J(y)$$

and

$$\sum_p (f^{\Delta} K^{\cdot})^p = \coprod_{x \in X} J(x) \quad .$$

We must show that for each $x \in X$ and each $a \in \Gamma(J(x))$ that $\mathrm{Tr}_f(da) = d(\mathrm{Tr}_f a)$, where d denotes the operator in the complex $K^{\cdot}$ (resp. $f^{\Delta} K^{\cdot}$).

Case 1. Suppose that x is closed in its fibre. Then Tr_f maps $f_*(J(x))$ into $J(y)$ where $y = f(x)$ (see the proof of [VI 4.4]), and we must show that the following diagram is commutative:

$$\begin{array}{ccc}
f_*(J(x)) & \xrightarrow{\quad d \quad} & \coprod\limits_{x \to x'} f_*(J(x')) \\
\Big\downarrow{\scriptstyle Tr_f} & & \Big\downarrow{\scriptstyle Tr_f} \\
J(y) & \xrightarrow{\quad d \quad} & \coprod\limits_{y \to y'} J(y')
\end{array} \quad ,$$

where we take those x' which are immediate specializations of x, and those y' which are immediate specializations of y. Clearly it is enough to consider each y' separately, so we must show the commutativity of the diagram

$$\begin{array}{ccc}
f_*(J(x)) & \xrightarrow{\quad d \quad} & \coprod\limits_{\substack{x \to x' \\ f(x')=y'}} f_*(J(x')) \\
\Big\downarrow{\scriptstyle Tr_f} & & \Big\downarrow{\scriptstyle Tr_f} \\
J(y) & \xrightarrow{\quad d \quad} & J(y')
\end{array} \quad .$$

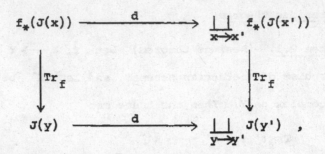

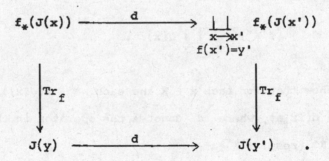

Given a particular element $a \in \Gamma(J(x))$, we can put a subscheme structure on $Z = \{x\}^-$ so that $a \in \Gamma(i_* i^\Delta f^\Delta K^\cdot)$ where $i\colon Z \longrightarrow X$ is the inclusion. Let $g\colon Z \longrightarrow Y$ be the composition fi. Then since i is a finite morphism $Tr_i = \rho_i$ is a morphism of complexes (see [VI §4]) and so using TRA 1 it will be sufficient to show that Tr_g is a morphism of complexes. In other words, we may replace X by Z and thus reduce to the case X is irreducible and x is the generic point.

Now by reason of codimensions [VI 3.4] each x' is closed in its fibre, i.e., the fibre of y' is discrete. Also f is a proper morphism, so by [EGA III 4.4.11] there is an open neighborhood U of y' such that the restriction of f to $f^{-1}(U)$ is a finite morphism. Thus we may assume that f itself is finite, in which case $Tr_f = \rho_f$ which is a morphism of complexes, and we are done.

Case 2. Suppose that x is not closed in its fibre. Then Tr_f maps $f_*(J(x))$ to zero. If no immediate specialization of x is closed in its fibre, then also $Tr_f \cdot d$ is zero on $f_*(J(x))$, and there is nothing to prove. So suppose that some immediate specializations x' of x are closed in their fibres. Then we have $f(x) = f(x') = y$ for all such x', and we must check that the diagram

$$\begin{array}{ccc} f_*(J(x)) & \xrightarrow{\ d\ } & \coprod\limits_{x \to x'} f_*(J(x')) \\ \Big\downarrow{\scriptstyle Tr_f} & & \Big\downarrow{\scriptstyle Tr_f} \\ 0 & \xrightarrow{\ d\ } & J(y) \end{array}$$

is commutative. As above, we fix an element $a \in \Gamma(J(x))$, and then can replace X by $\{x\}^-$ with a suitable subscheme structure. Thus we reduce to the case X irreducible and x its generic point.

Next we make the base extension $\operatorname{Spec} \mathcal{O}_y \longrightarrow Y$, and thus reduce to the case y is a closed point of Y, since everything is compatible with localization. Now since X is noetherian, and f maps the generic point of X to the closed point of Y, f factors through a closed subscheme Y' of Y defined by a suitable power of the maximal ideal of $\mathcal{O}_y$. Since $i\colon Y' \longrightarrow Y$ is a finite morphism, its trace is a morphism of complexes, so we can replace Y by Y', and so reduce to the case $Y = \operatorname{Spec} A$ with A a local Artin ring, and X irreducible of dimension one over Y.

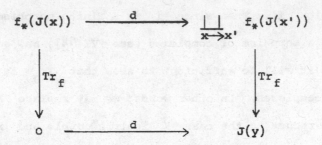

We refer to [EGA IV §22,25] for the following two results:

a) A proper scheme of dimension 1 over an Artin ring is projective, and

b) A projective scheme X of relative dimension $\leq n$ over a local ring A admits a finite morphism into $\mathbb{P}^n_A$.

(The adventurous reader can replace the reference by a proof of his own, using [EGA III 2.6.2] and [EGA III 4.7.1]. For a) one reduces to the case of a non-singular curve over a field, where any positive divisor of sufficiently high degree is very ample. For b) one puts X first in a large projective space $\mathbb{P}^N_A$, and then projects successively into smaller projective spaces.)

Using these results, we see that X is projective over Y, and admits a finite morphism onto $\mathbb{P}^1_A$. Since the theorem is known for a finite morphism, we reduce to the case $X = \mathbb{P}^1_A$, with A an artin ring. We now invoke [VI 5.4] and [VI 5.6] to reduce to the case where the residue field of A is algebraically closed. (Note that the base extension $Y' \longrightarrow Y$ is faithfully flat, so that it is enough to prove the theorem in the extended situation.) But this is Corollary 1.6, so we are done. q.e.d.

§3. The duality theorem for proper morphisms.

In this section we prove the long-awaited duality theorem for a proper morphism of locally noetherian preschemes $f: X \longrightarrow Y$. We will suppose the existence of a residual complex $K^\cdot$ on Y. Then $K^\cdot$ and $f^\Delta K^\cdot$ give rise to pointwise dualizing functors D_Y and D_X on Y and X, respectively, and we express the duality theorem as

$$D_X(F^\cdot) \xrightarrow{\sim} D_Y(\underline{\underline{R}} f_* F^\cdot)$$

for $F^\cdot \in D_{qc}(X)$.

Before proving the theorem, we must show that the trace map Tr_f of [VI 4.2] agrees in the case of a finite or projective morphism with the trace maps Trf_f of [III 6.5] and Trp_f of [III 4.3].

So let $f: X \longrightarrow Y$ be a finite morphism of preschemes, and assume that Y is <u>noetherian</u> and has <u>finite Krull dimension</u>. Let $K^\cdot$ be a residual complex on Y. Then we construct an isomorphism

$$\alpha_f: \; Qf_* f^\Delta K^\cdot \xrightarrow{\sim} \underline{\underline{R}} f_* f^\flat Q K^\cdot$$

as follows:

$$Qf_*f^{\Delta}K^{\cdot} \xrightarrow{\xi_{f_*}} \underline{\underline{R}}f_*Qf^{\Delta}K^{\cdot} \xrightarrow{\psi_f} \underline{\underline{R}}f_*Qf^{Y}K^{\cdot} \longrightarrow$$

$$\xrightarrow{(1)} \underline{\underline{R}}f_*QEf^{\flat}QK^{\cdot} \xrightarrow{(2)} \underline{\underline{R}}f_*f^{\flat}QK^{\cdot} ,$$

where as usual, Q denotes the natural map from complexes to elements of the derived category; ξ_{f_*} is the isomorphism of [I.5.1]; ψ_f is the isomorphism of [VI 3.1c]; the map (1) is the definition of f^Y [VI, §2]; and (2) is the isomorphism of [VI 1.1c] (here is where we need the hypothesis that Y is noetherian of finite Krull dimension).

<u>Proposition 3.1</u>. Let $f\colon X \longrightarrow Y$ be a finite morphism, with Y noetherian of finite Krull dimension, and let $K^{\cdot}$ be a residual complex on Y. Then there is a commutative diagram

$$\begin{array}{ccc} Qf_*f^{\Delta}K^{\cdot} & \xrightarrow{\quad \alpha_f \quad} & \underline{\underline{R}}f_*f^{\flat}QK^{\cdot} \\ & {\scriptstyle Q\,Tr_f}\searrow \quad \swarrow{\scriptstyle Trf_f\,Q} & \\ & QK^{\cdot} & \end{array}$$

where Tr_f is the trace of [VI 4.2] and Trf_f is the trace of [III 6.5].

Proof. Follows immediately from TRA 2 [VI 4.2] and the definition of ρ_f [VI §4].

Now let Y be a noetherian prescheme of finite Krull dimension, and let $f\colon X = \mathbb{P}^n_Y \longrightarrow Y$ be the structural map of an n-dimensional projective space over Y. Let $K^\cdot$ be a residual complex on Y. Then we define an isomorphism

$$\beta_f\colon\ Qf_*f^{\Delta}K^\cdot \xrightarrow{\ \sim\ } \underline{\underline{R}}f_*f^{\#}QK^\cdot$$

similar to the map α_f above, using ξ_{f_*}, the map φ_f of [VI 3.1d], the definition of f^Z [VI §2], and [VI 1.1c].

<u>Proposition 3.2</u>. Let Y be a noetherian prescheme of finite Krull dimension, let $X = \mathbb{P}^n_Y$, let $f\colon X \longrightarrow Y$ be the projection, and let $K^\cdot$ be a residual complex on Y. Then there is a commutative diagram

$$\begin{array}{ccc} Qf_*f^{\Delta}K^\cdot & \xrightarrow{\ \beta_f\ } & \underline{\underline{R}}f_*f^{\#}QK^\cdot \\ & {\scriptstyle Q\,\mathrm{Tr}_f}\searrow \quad \swarrow{\scriptstyle \mathrm{Trp}_f\,Q} & \\ & QK^\cdot & \end{array}$$

where Tr_f is the trace of [VI 4.2], and Trp_f is the trace of [III 4.3].

<u>Proof</u>. Choose a section $s\colon Y \longrightarrow X$ of f, and consider the following diagram:

$$\begin{array}{ccccccc}
QK^{\bullet} & \xrightarrow[\text{[VI 3.1b]}]{Q\,c_{s,f}} & Qf_{*}s_{*}s^{\Delta}f^{\Delta}K^{\bullet} & \xrightarrow[\text{[VI 4.2]}]{Q\,Tr_{s}} & Qf_{*}f^{\Delta}K^{\bullet} & \xrightarrow[\text{[VI 4.2]}]{Q\,Tr_{f}} & QK^{\bullet} \\
\Big\downarrow{\scriptstyle id} & & \Big\downarrow{\scriptstyle \alpha_{s},\beta_{f}} & & \Big\downarrow{\scriptstyle \beta_{f}} & & \Big\downarrow{\scriptstyle id} \\
QK^{\bullet} & \xrightarrow[\text{[III 8.1]}]{\psi_{s,f}^{Q}} & \underline{\underline{R}}f_{*}\underline{\underline{R}}s_{*}s^{\flat}f^{\#}QK^{\bullet} & \xrightarrow[\text{[III 6.5]}]{Trf_{s}^{Q}} & \underline{\underline{R}}f_{*}f^{\#}QK^{\bullet} & \xrightarrow[\text{[III 4.3]}]{Trp_{f}^{Q}} & QK^{\bullet}
\end{array} ,$$

where the notations have the sources indicated, and the second vertical arrow is obtained by sandwiching α_s in the middle of the four isomorphisms which define β_f. Now we make the following observations:

1). The composition of the upper row of arrows is the identity on $QK^{\bullet}$. This follows from TRA 1 [VI 4.2].

2). The composition of the lower row of arrows is also the identity on $QK^{\bullet}$. This is the statement of [III 10.1].

3). The left-hand square is commutative. This follows from VAR 5 [VI 3.1] and the definition of the isomorphisms α_s, β_f and (IV) of [VI §2].

4). The middle square is commutative. This is Proposition 3.1 above.

5). $\psi_{s,f}$ and Trp_f are isomorphisms by construction, hence Trf_s is an isomorphism.

6). $c_{s,f}$ is an isomorphism, hence $Q\ Tr_s$ and $Q\ Tr_f$ are isomorphisms. (Note incidentally we have used Theorem 2.1 above that Tr_f is a morphism of complexes, in order to consider $Q\ Tr_f$ in the first place.)

7). We conclude finally that the right-hand square is a commutative diagram of isomorphisms. q.e.d.

Now we come to the duality theorem itself. Let $f: X \longrightarrow Y$ be a <u>proper</u> morphism of noetherian preschemes of finite Krull dimension. Let $K^{\bullet}$ be a residual complex on Y. (Note that the existence of a residual complex imposes a slight restriction on the preschemes considered [VI 1.1] and [V §10].) We denote by D_Y (resp. $\underline{D}_Y$) the functor $\underline{\underline{R}}\ Hom^{\bullet}_Y(\cdot, QK^{\bullet})$ (resp. $\underline{\underline{R}}\ \underline{Hom}^{\bullet}(\cdot, QK^{\bullet})$), and by D_X (resp. $\underline{D}_X$) the functor $\underline{\underline{R}}\ Hom^{\bullet}_X(\cdot, Qf^{\Delta}K^{\bullet})$ (resp. $\underline{\underline{R}}\ \underline{Hom}^{\bullet}(\cdot, Qf^{\Delta}K^{\bullet})$). Then composing the morphism of [II.5.5] with ξf_* and $Q\ Tr_f$, we obtain the <u>duality morphism</u>

$$\theta_f:\ \underline{\underline{R}}f_*\ \underline{D}_X(F^{\bullet}) \longrightarrow \underline{D}_Y(\underline{\underline{R}}f_*F^{\bullet})$$

for $F^\cdot \in D^-(X)$. Applying the functor $\underline{\underline{R}}\Gamma(Y,\cdot)$ to both sides, and using [II 5.2] and [II 5.3] we obtain a global duality morphism

$$\Theta_f: \; D_X(F^\cdot) \longrightarrow D_Y(\underline{\underline{R}}f_*F^\cdot) \; .$$

Taking the cohomology of this, we get morphisms

$$\Theta_f^i: \; \mathrm{Ext}_X^i(F^\cdot, Qf^\Delta K^\cdot) \longrightarrow \mathrm{Ext}_Y^i(\underline{\underline{R}}f_*F^\cdot, QK^\cdot) \; .$$

Theorem 3.3 (Duality Theorem). Let $f: X \longrightarrow Y$ be a proper morphism of noetherian preschemes of finite Krull dimension, and let $K^\cdot$ be a residual complex on Y. Then the duality morphisms $\underline{\Theta}_f$, Θ_f, and Θ_f^i defined above are isomorphisms for all $F^\cdot \in D_{qc}^-(X)$.

(Note that the hypothesis of finite Krull dimension is needed only for the definition of $\underline{\Theta}_f$ (cf. [II 5.5]). If we restrict to bounded complexes $F^\cdot \in D_{qc}^b(X)$, we can state the theorem assuming only that X and Y are locally noetherian, and that the fibres of f are of bounded dimension. The proof is the same.)

Proof. We proceed in several steps, eventually using Chow's lemma to reduce to the case of projective space which we know already.

a) Clearly it is sufficient to show that $\underline{\theta}_f$ is an isomorphism. The question is local on Y, so we may assume Y is the spectrum of a local ring. In particular, we may assume that Y is noetherian, affine, and of finite Krull dimension. Using the lemma on way-out functors, we reduce to the case of a single quasi-coherent sheaf F on X.

b) Any quasi-coherent sheaf on F is the direct limit of its coherent subsheaves. In particular, it is a quotient of a direct sum of coherent sheaves. Thus using the lemma on way-out functors again, we reduce to the case of a direct sum of coherent sheaves. Now since $\underline{\underline{R}}\underline{Hom}^{\bullet}$ transforms direct sums in the first variable to direct products, we reduce to the case of a single coherent sheaf F on X.

c) Now since Y is affine, and all the sheaves considered are quasi-coherent, it is enough to show θ_f is an isomorphism.

d) If $f: X \longrightarrow Y$ and $g: Y \longrightarrow Z$ are two proper morphisms, and $K_Z^{\bullet}$ a residual complex on Z, and if we take $K^{\bullet} = g^{\Delta}K_Z^{\bullet}$, then for any $F^{\bullet} \in D_c^-(X)$ we have a commutative diagram

$$\begin{array}{ccc} D_X(F^{\bullet}) & \xrightarrow{\theta_{gf}} & D_Z(\underline{\underline{R}}g_* Rf_* F^{\bullet}) \\ & {}_{\theta_f}\searrow \qquad \nearrow_{\theta_g} & \\ & D_Y(\underline{\underline{R}}f_* F^{\bullet}) & \end{array}$$

.

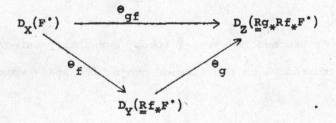

We deduce the following elementary but essential consequences:

(i) If Θ_f and Θ_g are isomorphisms for all arguments, so is Θ_{gf} .

(ii) If Θ_g and Θ_{gf} are isomorphisms for all arguments, so is Θ_f .

(iii) If Θ_f and Θ_{gf} are isomorphisms for all arguments, then Θ_g is an isomorphism for every complex of the form $\underline{R}f_*F^{\cdot}$ with $F^{\cdot} \in D_c^{-}(X)$.

e) By noetherian induction on X, we may assume the theorem proven for every morphism of the form $g: Z \longrightarrow Y$, where $g = fi$, and $i: Z \longrightarrow X$ is a closed immersion of Z onto a subscheme of X, different from X.

f) We now apply Chow's Lemma [EGA II 5.6.1] to deduce the existence of a scheme X' projective over Y, together with a projective Y-morphism $g: X' \longrightarrow X$, which is an isomorphism on a non-empty open subset U of X (we may assume $X \neq \emptyset$!). Given a coherent sheaf F on X, consider the natural map

$$\begin{array}{ccc} X' & \xrightarrow{\ g\ } & X \\ & {\scriptstyle fg}\searrow & \downarrow{\scriptstyle f} \\ & & Y \end{array}$$

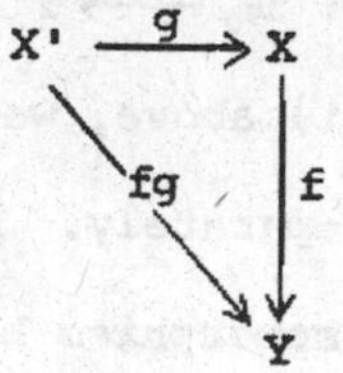

$u\colon F \longrightarrow \underline{\underline{R}}g_* g^* F$, and embed it in a triangle

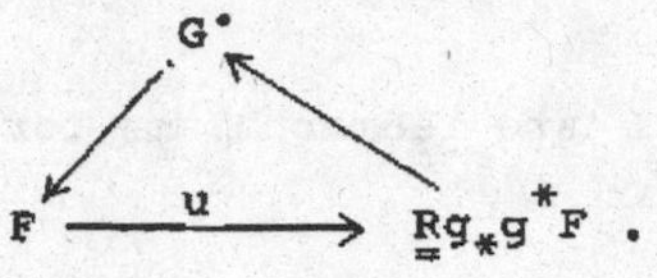

Then since g is an isomorphism on U, $G^{\cdot}$ has support on $X-U$, and so $\theta_f(G^{\cdot})$ is an isomorphism by our induction hypothesis e). Thus it will be sufficient to show that $\theta_f(\underline{\underline{R}}g_* g^* F)$ is an isomorphism. Using (iii) above, we reduce to showing θ_g and θ_{fg} are isomorphisms for all arguments, and so we reduce to the case f is projective.

g) Since Y is affine, we can embed X in a suitable projective space over Y, say $f = pi$, with $p\colon \mathbb{P}^n_Y \longrightarrow Y$ the projection. Using (i) above, we can treat i and p separately. Now i is a finite morphism, so θ_i is an isomorphism by Proposition 3.1 above, and [III 6.7]. Also θ_p is an isomorphism by Proposition 3.2 above, and [III 5.1].

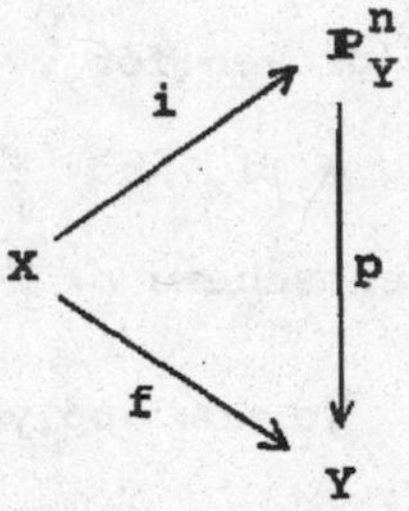

q.e.d.

We can now pull ourselves up by our bootstraps, and obtain a theory of $f^!$ and Tr_f for complexes with <u>coherent cohomology</u> and schemes <u>admitting dualizing complexes</u>.

<u>Corollary 3.4</u>. We consider the category of noetherian preschemes which admit a dualizing complex, and we consider morphisms of finite type. Then

(a). For every such morphism $f: X \longrightarrow Y$, there is a theory of $f^!$ consisting of a functor

$$f^! : D_c^+(Y) \longrightarrow D_c^+(X)$$

plus the data 2)-5) and properties VAR 1 - VAR 6 of [III 8.7] (only leave out the word "embeddable" wherever it occurs).

(b). For every such <u>proper</u> morphism $f: X \longrightarrow Y$, there is a theory of trace consisting of a functorial morphism

$$Tr_f: \underline{\underline{R}}f_* f^! \longrightarrow 1$$

with the properties TRA 1 - TRA 4 of [III 10.5] (only leave out the phrase "projectively embeddable" wherever it occurs).

(c). For every such proper morphism $f: X \longrightarrow Y$, the duality morphism

$$\underline{\theta}_f: \ \underline{\underline{R}}f_* \underline{\underline{R}}\underline{Hom}^{\bullet}_X(F^{\bullet}, f^! G^{\bullet}) \longrightarrow \underline{\underline{R}} \ \underline{Hom}^{\bullet}_Y(\underline{\underline{R}}f_* F^{\bullet}, G^{\bullet}) ,$$

obtained by composing the morphism of [II 5.5] with Tr_f in the second place, is an isomorphism for $F^{\bullet} \in D^-_{qc}(X)$ and $G^{\bullet} \in D^+_c(Y)$. (Compare [III 11.1].)

Proof. (a) Let $K^{\bullet}$ be a residual complex on Y (and observe that $QK^{\bullet}$ is then a dualizing complex on Y, and $Qf^{\Delta}K^{\bullet}$ a dualizing complex on X). Let $\underline{D}_Y$ and $\underline{D}_X$ be as in the theorem above, and define

$$f^!(G^{\bullet}) = \underline{D}_X(\underline{\underline{L}}f^* \underline{D}_Y(G^{\bullet}))$$

for all $G^{\bullet} \in D^+_c(Y)$. Note that $\underline{D}_Y(G^{\bullet}) \in D^-_c(Y)$, so that $\underline{\underline{L}}f^*$ of it is defined. Here is where we need that $QK^{\bullet}$ is a dualizing complex, not just pointwise dualizing. Observe also that $f^!$ is independent of the choice of $K^{\bullet}$ (use [V 3.1]).

The construction of the isomorphisms $c_{f,g}$ is easy, using [VI 3.1]. For the isomorphisms d_f and e_f, we use [III 6.9b] and [III 2.4], respectively. The details of these constructions, and verifications of VAR 1 - VAR 6 are left to the reader (!).

Observe, by the way, that this part of the Corollary does not depend on the duality theorem, and could have come just after the construction of f^{Δ} [VI 3.1].

(b) To define Tr_f, let $G^\bullet \in D_c^+(Y)$, and use [II 5.10], [VI 4.2], Theorem 2.1 above, and [V 2.1]:

$$\begin{array}{ccc}
\underline{\underline{R}}f_* f^! G^\bullet & = & \underline{\underline{R}}f_* \, \underline{\underline{R}} \, \underline{\mathrm{Hom}}^\bullet_X(\underline{\underline{L}}f^* \underline{D}_Y(G^\bullet), \, Q f^\Delta K^\bullet) \\
\vdots & & \downarrow {\scriptstyle [\mathrm{II}\ 5.10]} \\
\vdots & & \underline{\underline{R}} \, \underline{\mathrm{Hom}}^\bullet_Y(\underline{D}_Y(G^\bullet), \, \underline{\underline{R}}f_* Q f^\Delta K^\bullet) \\
\vdots & & \downarrow {\scriptstyle Q\,\mathrm{Tr}_f} \\
\downarrow & & \\
G^\bullet & \xrightarrow{\ \sim\ } & \underline{\underline{R}} \, \underline{\mathrm{Hom}}^\bullet_Y(\underline{D}_Y(G^\bullet), \, QK^\bullet)
\end{array} \quad .$$

The verification of TRA 1 is clear, using TRA 1 of [VI 4.2]. For TRA 2 we use [III 6.9c]. Details left to reader (!).

(c) To prove the duality formula, we reduce as in part b) of the proof of the theorem above, to the case of complexes $F^\bullet$ with <u>coherent</u> cohomology. So let $F^\bullet \in D_c^-(X)$ and $G^\bullet \in D_c^+(Y)$. Consider

$$\underline{\underline{R}}f_* \, \underline{\underline{R}} \, \underline{\mathrm{Hom}}_X(F^\bullet, f^! G^\bullet) \ . \qquad (1)$$

We write $F^\bullet = \underline{D}_X \underline{D}_X(F^\bullet)$ and $f^! G^\bullet = \underline{D}_X(\underline{\underline{L}}f^* \underline{D}_Y(G^\bullet))$. Now since $\underline{D}_X$ is a dualizing functor, it transforms $\underline{\underline{R}}$ <u>Hom</u> into $\underline{\underline{R}}$ <u>Hom</u> of the duals of the arguments, with the order reversed. Thus (1) becomes

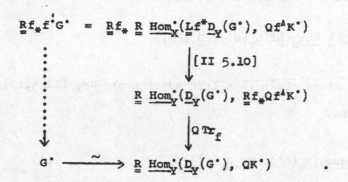

$$\underline{\underline{R}}f_* \, \underline{D}_X \, \underline{\underline{R}} \, \underline{\underline{Hom}}_X(\underline{\underline{L}}f^*\underline{D}_Y(G^\cdot), \underline{D}_X(F^\cdot)) \, . \tag{2}$$

Applying the duality theorem above, this becomes

$$\underline{D}_Y \, \underline{\underline{R}}f_* \, \underline{\underline{R}} \, \underline{\underline{Hom}}_X(\underline{\underline{L}}f^*\underline{D}_Y(G^\cdot), \underline{D}_X(F^\cdot)) \tag{3}$$

which in turn, since $\underline{\underline{L}}f^*$ is a left adjoint of $\underline{\underline{R}}f_*$ [II 5.10], is isomorphic to

$$\underline{D}_Y \, \underline{\underline{R}} \, \underline{\underline{Hom}}_Y(\underline{D}_Y(G^\cdot), \underline{\underline{R}}f_*\underline{D}_X(F^\cdot)) \, . \tag{4}$$

Now applying duality to the second argument, (4) becomes

$$\underline{D}_Y \, \underline{\underline{R}} \, \underline{\underline{Hom}}_Y(\underline{D}_Y(G^\cdot), \underline{D}_Y(\underline{\underline{R}}f_*F^\cdot))$$

and since $\underline{D}_Y$ is dualizing, this is

$$\underline{\underline{R}} \, \underline{\underline{Hom}}_Y(\underline{\underline{R}}f_*F^\cdot, G^\cdot) \, .$$

The reader may check that our chain of isomorphisms is indeed $\underline{\theta}_f$, which proves (c).

<u>Proposition 3.5</u> (Compatibility of Local and Global duality). Let $f: X \longrightarrow Y$ be a proper morphism of noetherian schemes, and let $K^\cdot$ be a residual complex on Y. Let $x \in X$ be a closed point, with local ring $A = \mathcal{O}_x$. Let $R^\cdot$ be the dualizing complex $(Qf^\Delta K^\cdot)_x$ on A, and assume that $R^\cdot$ is

normalized [V §6]. Let $I = \Gamma_x(R^\cdot)$. Let $F^\cdot \in D_c^+(X)$. Then the diagram

$$\begin{array}{ccc}
\underline{\underline{R}}\Gamma_x(F^\cdot) & \xrightarrow{\;\theta_x\;} & \mathrm{Hom}_A^\cdot(\mathrm{Hom}_A^\cdot(F_x^\cdot,R^\cdot),\ I) \\
\downarrow{\scriptstyle\alpha} & & \downarrow{\scriptstyle\beta} \\
\underline{\underline{R}}f_*(F^\cdot) & \xrightarrow{\;\underline{D}_Y(\theta_f)\;} & \underline{\mathrm{Hom}}_Y^\cdot(\underline{\underline{R}}f_*\ \underline{\mathrm{Hom}}_X^\cdot(F^\cdot,Qf^\Delta K^\cdot),QK^\cdot)
\end{array}$$

is commutative, where θ_x is the local duality isomorphism [V 6.2], α is the natural map of derived functors obtained from the inclusion $\Gamma_x \subseteq f_*$, β is obtained from the stalk map

$$\underline{\underline{R}}f_*\ \underline{\mathrm{Hom}}_X^\cdot(F^\cdot,Qf^\Delta K^\cdot) \longrightarrow \mathrm{Hom}_A^\cdot(F_x^\cdot,R^\cdot)$$

and the trace map

$$\mathrm{Tr}_{f,x}\colon\ f_*J(x) = I \longrightarrow K^\cdot ,$$

and $\underline{D}_Y\theta_f$ is the transpose by $\underline{D}_Y$ of the global duality isomorphism of Theorem 3.3. (Note we write four times $\mathrm{Hom}^\cdot$ instead of $\underline{\underline{R}}\ \mathrm{Hom}^\cdot$, because the second argument in each case is injective.)

Proof. Immediate from the definitions of the maps in question.

Remark. This compatibility is the one needed to complete the proof of "Lichtenbaum's theorem" [LC theorem 6.9, see parenthetical remark in middle of p. 103].

§4. Smooth morphisms.

In this section we give the special case of the duality theorem for a proper smooth morphism of locally noetherian preschemes. In this case we can eliminate the hypothesis that our preschemes admit residual complexes.

The results below are valid practically without change for Cohen-Macaulay morphisms (see [V.9.7]). In that case one defines $\omega_{X/Y}$ to be the unique cohomology group of $f^!(\mathcal{O}_Y)$. The functor $f^!$ is defined only locally, but one can glue the sheaves $\omega_{X/Y}$ to obtain a global one. One then defines $f^{\#} = f^{*} \otimes \omega_{X/Y}[n]$ as in the smooth case. We leave the details to the reader.

Throughout this section, $f\colon X \longrightarrow Y$ will be a proper, smooth morphism of locally noetherian preschemes, and we will suppose for simplicity that the fibres are all of the same dimension, say n. We denote by $\omega_{X/Y}$ the sheaf $\Omega^n_{X/Y}$ of relative n-differentials, and by $f^{\#}$ the functor

$$f^{\#}\colon D(Y) \longrightarrow D(X)$$

given by

$$f^{\#}(G^{\cdot}) = f^{*}(G^{\cdot}) \otimes \omega_{X/Y}[n]$$

(compare [III, §§1,2]).

<u>Theorem 4.1</u>. For every proper, smooth morphism $f\colon X \longrightarrow Y$ of relative dimension n of locally noetherian preschemes, there is a morphism

$$\gamma_f\colon\ R^n f_*(\omega_{X/Y}) \longrightarrow \mathcal{O}_Y$$

with the properties b)-g) of [III 11.2].

<u>Proof</u>. We will first consider the case where Y is noetherian and admits a dualizing complex. Then by Corollary 3.4, we have

$$f^!(\mathcal{O}_Y) = f^{\#}(\mathcal{O}_Y) = \omega_{X/Y}[n] ,$$

and we have a trace map

$$\mathrm{Tr}_f\colon\ \underline{\underline{R}}f_* f^!(\mathcal{O}_Y) \longrightarrow \mathcal{O}_Y .$$

Taking the cohomology in degree n, we obtain a map

$$\gamma_f\colon\ R^n f_*(\omega_{X/Y}) \longrightarrow \mathcal{O}_Y$$

as required. The proofs of the properties b)-g) are similar to loc. cit. except for c), which we will leave to the reader.

For the general case, the question is local by c), so we may assume $Y = \operatorname{Spec} A$ is the spectrum of a noetherian ring A. We consider flat base extensions $u_i\colon\ Y_i \longrightarrow Y$, where Y_i is the spectrum of a complete local noetherian ring B_i, flat over A. By the Cohen structure theorem [14, (31.1)], each B_i is a quotient of a regular local ring, hence admits a dualizing complex, and so the theorem holds for Y_i. Thus for each i we have a morphism

$$\gamma_{f_i}: R^n f_{i*}(\omega_{X_i/Y_i}) \longrightarrow \mathcal{O}_{Y_i} .$$

Furthermore, by c), if $v_{ij}: Y_i \longrightarrow Y_j$ is a morphism compatible with the morphisms u_i, u_j, we have

$$\gamma_{f_i} = v_{ij}^* \gamma_{f_j} .$$

By the lemma below, applied to the A-module

$$\mathrm{Hom}_Y(R^n f_*(\omega_{X/Y}), \mathcal{O}_Y) ,$$

there is a unique

$$\gamma_f: R^n f_*(\omega_{X/Y}) \longrightarrow \mathcal{O}_Y$$

such that $\gamma_{f_i} = u_i^* \gamma_f$ for all i. By virtue of its construction, this γ_f has the properties b)-g), which completes the proof (details left to reader!).

<u>Lemma 4.2</u>. Let A be a noetherian ring, and let M be an A-module. We consider the category $(B_i)_{i\in I}$ of A-algebras which are complete noetherian local rings, flat over A, and morphisms of A-algebras. Then the natural map

$$\varphi\colon M \longrightarrow \varprojlim M \otimes_A B_i$$

is injective, and if moreover M is of finite type, φ is bijective.

<u>Proof</u>. To see that φ is injective, it is sufficient to notice that the map

$$M \longrightarrow \prod M \otimes \hat{A}_{\mathfrak{m}}$$

is injective, where the product is taken over the maximal ideals $\mathfrak{m}$ of A.

For surjectivity, suppose first that $M \cong A/\mathfrak{p}$, where $\mathfrak{p}$ is a prime ideal of A. Then we have the following commutative diagram:

$$\begin{array}{ccccccc}
0 \longrightarrow & M & \longrightarrow & M_{\mathfrak{p}} & \longrightarrow & M_{\mathfrak{p}}/M & \longrightarrow 0 \\
 & \Big\downarrow{\scriptstyle \varphi_1} & & \Big\downarrow{\scriptstyle \varphi_2} & & \Big\downarrow{\scriptstyle \varphi_3} & \\
0 \longrightarrow & \varprojlim M \otimes B_i & \longrightarrow & \varprojlim M_{\mathfrak{p}} \otimes B_i & \longrightarrow & \varprojlim (M_{\mathfrak{p}}/M) \otimes B_i & .
\end{array}$$

Now φ_2 is bijective, because $M_{\mathfrak{p}} = k(\mathfrak{p})$, and so the natural map

$$M_{\mathfrak{p}} \longrightarrow M_{\mathfrak{p}} \otimes_A (A_{\mathfrak{p}})^\wedge$$

is bijective. On the other hand, φ_3 is injective, so φ_1 is bijective.

Now let M be an arbitrary A-module of finite type. Then we can find a filtration

$$0 = M_0 \subseteq M_1 \subseteq \ldots \subseteq M_r = M$$

whose quotients M_i/M_{i-1} are of the form $A/\mathfrak{p}_i$ with $\mathfrak{p}_i \in \operatorname{Ass} M$. By using the 5-lemma and induction on r, we reduce to the previous case.

Remark. One sees from the proof that it would be sufficient to consider only B_i of the form $(A_{\mathfrak{p}})^\wedge$ where $\mathfrak{p}$ is either maximal, or an element of Ass M.

Corollary 4.3. a) For every proper, smooth morphism $f: X \longrightarrow Y$ of noetherian preschemes of finite Krull dimension. There is a trace morphism

$$\operatorname{Tr}_f : \underline{\underline{R}}f_* f^{\#} G^{\cdot} \longrightarrow G^{\cdot}$$

for $G^\cdot \in D^b_{qc}(Y)$, satisfying **TRA 1-TRA 4** of [III 10.5] (but where TRA 4 is valid for arbitrary base extension).

b) The resulting duality morphism

$$\underline{\theta}_f : \underline{\underline{R}}f_* \,\underline{\underline{R}}\, \underline{Hom}^\cdot_X(F^\cdot, f^\# G^\cdot) \longrightarrow \underline{\underline{R}}\, \underline{Hom}^\cdot_Y(\underline{\underline{R}}f_* F^\cdot, G^\cdot)$$

is an isomorphism for $F^\cdot \in D^-_{qc}(X)$ and $G^\cdot \in D^b_{qc}(Y)$.

<u>Proof</u>. Define Tr_f by the projection formula and γ_f. Details left to reader!

Index of Definitions

Index of Notations

Of course there is the usual collection of variable notations: A,B for abelian categories or rings, F,G for functors or sheaves, X,Y for preschemes, x,y for points of them, f,g for morphisms of preschemes, φ,ψ for morphisms of functors on sheaves, etc. In general no underline denotes a group, as Hom(F,G), one underline denotes a sheaf, as $\underline{\mathrm{Hom}}(F,G)$ or $\underline{\mathrm{Tor}}(F,G)$, two underlines denotes an object in a derived category, as $\underline{\underline{R}}\,\mathrm{Hom}^{\bullet}(F^{\bullet},G^{\bullet})$, or $F^{\bullet}\underline{\underline{\otimes}}G^{\bullet}$, and a dot denotes a complex.

In the list of stable notations below, we have distinguished five categories: Latin alphabet, other alphabets, superscripts, subscripts, and arbitrary symbols.

Latin alphabet

TR1-TR4	axioms of a triangulated category	I.1
TRA	axioms of trace	III 10.5, VI 4.2
tr. d.	transcendence degree	
Trf	trace for a finite morphism	III.6
Trp	trace for a projective morphism	III.4
VAR	axioms for $f^{\cdot}$ or f^{Δ}	III 8.7, VI 3.1
Z	cycles of a complex	

Greek and other alphabets

γ	trace for smooth morphism	III 3.4, III 11.2, VII 4.1
Γ	global sections of a sheaf	
ζ	change of differentials	III 1.5
η	fundamental local isomorphism	III 7.3
θ	duality morphism	III.5, III.6, III.11, V.6, VII.3
ξ	map of a functor to its derived functor	I.5
$\sigma_{>n}$	truncated complex	I.7
$\tau_{>n}$	truncated complex	I.7
ψ	residue isomorphism	III.8
ω	sheaf of differentials	III.1
Ω	sheaf of differentials	III.1
$\mathcal{O}_X$	structure sheaf of a prescheme	
$\mathfrak{m}$	maximal ideal of a ring	
$\mathfrak{p}$	prime ideal of a ring	
$\mathbb{A}$	affine space	
$\mathbb{P}$	projective space	
$\mathbb{Q}$	the rational numbers	
$\mathbb{Z}$	the integers	

Subscripts

A'	as in $K_{A'}(A)$, $D_{A'}(A)$	I.4
c	as in $D_c(X)$	II.1
fid	as in $D(X)_{fid}$	I.7
fTd	as in $D(X)_{fTd}$	II.4

Gor(Z')	as in $D(X)_{Gor(Z^{\cdot})}$	IV.3
$\mathfrak{p}$	as in $A_{\mathfrak{p}}, M_{\mathfrak{p}}$: localization at a prime ideal	
qc	as in $D_{qc}(X)$	II.1
x	as in $\mathcal{O}_x$: local ring at a point x	
	as in F_x: stalk of a sheaf at x	
	as in H^i_x: local cohomology	IV.1
Z	as in Γ_Z, H^i_Z	IV.1
I,II	as in $\underline{\underline{R}}_I$, $\underline{\underline{R}}_{II}$	I.6
>n	as in $\sigma_{>n}$, $\tau_{>n}$	I.7
*	as in f_*: direct image	II.2

Superscripts

b	as in K^b, D^b: bounded	I.2,I.4
i	as in X^i: i^{th} term of a complex	
	as in H^i: i^{th} cohomology group	
	as in R^iF: i^{th} derived functor	
y	as in f^y	VI.2
z	as in f^z	VI.2
·	as in $F^{\cdot}, X^{\cdot}$: denotes a complex	
··	as in $C^{\cdot\cdot}$: double complex	I.7
°	as in $K(A)^{\circ}$: opposed category	
*	as in f^*: inverse image	II.4
+	as in $K^+, D^+, \underline{\underline{R}}^+$: bounded below	I.2,I.4,I.5
-	as in $K^-, D^-, \underline{\underline{R}}^-$: bounded above	I.2,I.4,I.5

!	as in $f^!$	Intr.,III.8,VII.3
♭	as in $f^\flat$	III.6
#	as in $f^\#$	III.2
Δ	as in f^Δ	VI.3
ˇ	as in $L^\vee$: dual sheaf	II.5.16,III.1
—	as in $\{x\}^-$: closure	
~	as in $\tilde{M}$: sheaf associated to a module	
^	as in $\hat{A}$: completion of local ring	

Arbitrary Symbols

$\mid$	restriction	
$>>$	enough larger than	
$\times$	product	
$\otimes$	tensor product	
$\oplus$	direct sum	
$\prod$	product	
$\coprod$	disjoint union or sum	
[ ]	shift operator	I.2
	closed interval	V.8
	reference to Bibliography or other chapter	Intro.
$\cup$	union	
{ }	the set of	
$\longrightarrow$	morphism	
$\longmapsto$	effect of a map on elements	
$\cdot\!\!\longrightarrow$	distinguished side of triangle	I.1
$\cdots\!>$	morphism being constructed	
$\cdot$	as in $F\cdot G$: composition of functors	

BIBLIOGRAPHY

GT M. Artin, "Grothendieck Topologies", mimeographed seminar notes, Harvard (1962).

1. H. Bass, "Injective dimension in noetherian rings", T.A.M.S. 102 (1962), 18-29.

2. _______, "On the ubiquity of Gorenstein rings", Math. Zeitschrift 82 (1963), 8-28.

M H. Cartan and S. Eilenberg, "Homological Algebra", Princeton University Press (1956).

3. C. Chevalley, "Introduction to the theory of algebraic functions of one variable", Amer. Math. Soc. Surveys (1951).

4. Eckmann and Schopf, "Über injektive Moduln", Archiv der Math. 4 (1953).

5. P. Gabriel, "Des Catégories Abéliennes", Bull. Soc. Math. Fr. 90 (1962), 323-448.

6. Giraud, thesis (to appear)

G R. Godement, "Topologie algébrique et théorie des faisceaux", Hermann, Paris (1958).

7. D. Gorenstein, "An arithmetic theory of adjoint plane curves", T.A.M.S. 72 (1952), 414-436.

T A. Grothendieck, "Sur quelque points d'algèbre homologique", Tohoku Math. J. IX (1957), 119-221.

8. ______________, "Théorèmes de dualité pour les faisceaux algébriques cohérents", seminaire Bourbaki, no. 149, Secr. Math. I.H.P. Paris (1957).

9. ______________, "The cohomology theory of abstract algebraic varieties" in Int. Cong. of Math. at Edinburgh, 1958, Cambridge Univ. Press (1960), 103-118.

EGA A. Grothendieck, "Eléments de Géométrie Algébrique", Publ. Math. I.H.E.S. Paris 4,8,11,17,20,24, 28 (1960 ff).

LC ______________, "Local Cohomology", mimeographed seminar notes by R. Hartshorne, Harvard (1961).

SGA ______________, "Séminaire de Géométrie Algébrique", notes polycopiés, I.H.E.S. Paris (1960-61; 1962).

10. ______________, "Résidus et dualité", pré-notes pour un "Séminaire Hartshorne" manuscript (1963).

11. R. Hartshorne, "A property of A-sequences", Bull. Soc. Math. Fr. 94 (1966).

12. S. Mac Lane, "Categorical Algebra", Colloquium lectures given at Boulder, Colo. Aug. 27-30 (1963) at the 68th summer meeting of the AMS.

13. E. Matlis, "Injective modules over noetherian rings", Pac. J. Math. 8 (1958), 511-528.

14. M. Nagata, "Local Rings", Interscience Tracts no. 13, J. Wiley and Sons, N.Y. (1962).

15. J.-P. Serre, "Un théorème de dualité", Comment. Math. Helvet. 29 (1955), 9-26.

FAC __________, "Faisceaux algébriques cohérents", Annals of Math. 61 (1955), 197-278.

16. __________, "Groupes algébriques et corps de classes", Paris, Hermann (1959).

17. J.-L. Verdier, "Théorème de dualité de Poincaré", Comptes Rendus 256 (1963), 2084-2086.

18. ____________, thesis (to appear)

19. O. Zariski, "Complete linear systems on normal varieties and a generalization of a lemma of Enriques-Severi", Annals of Math. 55 (1952), 552-592.

20. __________, "Algebraic sheaf theory", Scientific Report on the second summer institute, Part III, Bull. Amer. Math. Soc. 62 (1956), 117-141.

ZS O. Zariski and P. Samuel, "Commutative Algebra", 2 vols., van Nostrand, Princeton (1958,1960).

21. M.F. Atiyah and R. Bott, "A Lefschetz fixed point formula for elliptic differential operators", Bull. Amer. Math. Soc. 72 (1966), 245-250.

22. A. Grothendieck, "De Rham cohomology of algebraic varieties", to appear, Publ. Math. I.H.E.S.

APPENDIX : COHOMOLOGIE A SUPPORT PROPRE ET CONSTRUCTION DU FONCTEUR $f^!$.

par P. DELIGNE (1)

Verdier a montré que dans le cas topologique, le formalisme de la dualité de Poincaré se ramenait à des problèmes locaux en haut (voir [1]). Pour transposer sa construction au cadre schématique, il faut disposer d'une théorie de la cohomologie "à support propre" pour les faisceaux cohérents.

Sauf mention explicite du contraire, tous les préschémas considérés sont noethériens et les préfaisceaux quasi-cohérents.

n° 1. Le sorite des pro-objets.

__Proposition 1__. __Soit__ C __une__ U -__catégorie__ (2) __où existent les__

(1) Ceci est une version complétée d'une lettre de P. Deligne à R. Hartshorne (lettre du 3 Mars 1966). Les notes de bas de page ont été ajoutées par le copiste.

(2) U désigne un univers fixé dans toute la suite.

<u>limites inductives finies</u>. <u>Soit</u> h <u>un foncteur</u> $C^o \longrightarrow$ (ens). <u>Les conditions suivantes sont équivalentes</u> :

(i) h <u>est limite inductive, selon un petit</u> (3) <u>ensemble ordonné filtrant, de foncteurs représentables</u>.

(ii) h <u>est limite inductive, selon une petite catégorie filtrante, de foncteurs représentables</u>.

(iii) h <u>transforme</u> $\varinjlim$ <u>finies en</u> $\varprojlim$ <u>finies, et il existe un petit ensemble d'objets tel que tout élément d'un</u> h(X) <u>se factorise par l'un d'eux</u>.

Les implications (i) $\Longrightarrow$ (ii) $\Longrightarrow$ (iii) sont triviales et (iii) $\Longrightarrow$ (ii) standard (h est limite des h_F selon la catégorie des (F,α) pour $\alpha \in h(F)$ et F dans la sous-catégorie de C stable par $\varinjlim$ finie engendrée par le petit ensemble donné). Reste à prouver que (ii) $\Longrightarrow$ (i). Si $\mathcal{L}$ et $\mathcal{M}$ sont deux catégories filtrantes, un foncteur $F : \mathcal{L} \longrightarrow \mathcal{M}$ est dit cofinal si $\varinjlim\limits_{\mathcal{L}} h_{F(L)}$ est le foncteur final. Pour tout $G : \mathcal{M} \to \mathcal{N}$, on a alors $\varinjlim\limits_{\mathcal{M}} G = \varinjlim\limits_{\mathcal{L}} GF$. Il s'agit de prouver:

<u>lemme</u>. <u>Pour toute petite catégorie filtrante</u> $\mathcal{L}$, <u>il existe un foncteur cofinal d'un petit ensemble ordonné filtrant dans</u> $\mathcal{L}$.

La première projection : $\mathcal{L} \times N \longrightarrow \mathcal{L}$, est un foncteur cofinal (N muni de l'ordre naturel) ; ceci permet de se

---

(3) "petit" = " $\in$ U" .

ramener à supposer que Ob$\mathcal{L}$, préordonné par Hom (X,Y) $\neq \emptyset$, n'a pas de plus grand élément. On ordonne par inclusion l'ensemble E des sous-catégories finies de $\mathcal{L}$ ayant un seul objet final (fini signifie d'ensemble de flèches fini). On définit un foncteur de E dans $\mathcal{L}$ en associant à chacune de ces catégories son objet final. Sous les hypothèses faites, E est filtrant et ce foncteur cofinal.

Les foncteurs vérifiant les conditions équivalentes de la proposition 1 sont les Ind -objets de C . Ils forment une sous-catégorie pleine Ind C de <u>Hom</u> (C°, (Ens)) , stable par limite inductive. C $\mapsto$ Ind C est un foncteur (cat)$\rightarrow$(cat). Si les limites inductives filtrantes existent dans C , pour tout h $\in$ Ind C , le foncteur X $\longrightarrow$ Hom (h, h_X) est coreprésentable, d'où un foncteur, noté $\varinjlim$, de Ind C dans C . En particulier, on a toujours un foncteur Ind Ind C $\longrightarrow$ Ind C , et tout foncteur C $\longrightarrow$ Ind $\mathcal{D}$ se prolonge en un foncteur Ind C $\longrightarrow$ Ind $\mathcal{D}$. Dans Ind C , les limites inductives filtrantes sont exactes, et seront notées "$\varinjlim$" ; en particulier, identifiant C à une sous-catégorie pleine de Ind C , si $(X_i)_{i \in I}$ est un système inductif filtrant dans C ,

$$\varinjlim_{i \in I} X_i = \varinjlim \text{"}\varinjlim_{i \in I}\text{"} X_i$$

Pour qu'un foncteur sur Ind C soit représentable, il faut et suffit qu'il transforme $\varinjlim$ finies en $\varprojlim$ finies, $\varinjlim$ filtrantes en $\varprojlim$ filtrantes et que sa restriction à C satisfasse à la condition de petitesse de la prop. 1 (iii) (en effet, la proposition 1 montre alors que sa restriction à C est un Ind -objet, auquel il est partout égal vu la condition sur les limites).

Ce qui précède, sauf la prop. 1 (iii) et l'assertion précédente reste vrai en utilisant partout des limites de suites (ou dénombrables, c'est la même chose).

On définit par dualité la catégorie pro C des pro-objets (sous-catégorie pleine de Hom (C, (Ens))°) .

Proposition 2. Soit X un préschéma quasi-compact quasi-séparé (non nécessairement noethérien). La catégorie des faisceaux quasi-cohérents sur X est équivalente à la catégorie des Ind - objets de la catégorie des faisceaux quasi-cohérents de présentation finie sur X .

La flèche est $\text{"}\varinjlim\text{"}\mathcal{F}_i \longmapsto \varinjlim \mathcal{F}_i$. Si $\mathcal{F}$ est de p.f. (présentation finie), pour tout système inductif filtrant $(\mathcal{G}_i)_{i\in I}$, $\mathrm{Hom}_X(\mathcal{F}, \varinjlim \mathcal{G}_i) = \varinjlim \mathrm{Hom}_X(\mathcal{F}, \mathcal{G}_i)$ (par un recollement fini qui commute à $\varinjlim$, on se ramène au cas affine). Le foncteur précédent est donc pleinement fidèle. Son image est stable par limites inductives filtrantes et sommes finies. Il suffit de prouver que pour tout $x\in X$, tout $\mathcal{F}$ sur X , et tout $s\in\mathcal{F}_x$, il existe $\mathcal{G}$ de p.f. et une flèche de $\mathcal{G}$ dans $\mathcal{F}$ dans l'image de laquelle se trouve s :
$\mathcal{F}$ sera limite d'images de faisceaux de p.f. et chacune d'elles quotient d'un faisceau de p.f. par une limite de sous-faisceaux

de type fini.

Soit <u>sur</u> un voisinage quasi-compact U de x une flèche f : $\mathcal{C} \longrightarrow \mathcal{F}$, $\mathcal{C}$ de p.f. , s dans l'image. Il suffit de prolonger $\mathcal{C}$ et f à X entier, et, procédant pas à pas, il suffit de savoir prolonger à $U \cup V$ si V est affine, ce qui revient à effectuer un prolongement de $U \cap V$ à V :

<u>Lemme</u>. <u>Soit</u> $\mathcal{C}$ <u>un faisceau de p.f. sur un ouvert quasi-compact</u> U <u>d'un schéma affine</u> X , $\mathcal{F}$ <u>un faisceau sur</u> X <u>et</u> f <u>une flèche de</u> $\mathcal{C}$ <u>dans</u> $\mathcal{F}|U$. <u>Il est possible de prolonger</u> $\mathcal{C}$ <u>et</u> f <u>en</u> $\overline{\mathcal{C}}$ <u>et</u> $\overline{f}$, <u>définis sur</u> X ($\overline{\mathcal{C}}$ étant de p.f.).

Soit ι l'inclusion de U dans X et $\mathcal{C}_1$ le produit fibré de $\iota_*\mathcal{C}$ et $\mathcal{F}$ sur $\iota_*\iota^*\mathcal{F}$. $\mathcal{C}_1$ prolonge $\mathcal{C}$ et s'envoie dans $\mathcal{F}$. Si on représente $\mathcal{C}_1$ comme limite de ses sous-faisceaux de type fini, comme U est quasi-compact et $\mathcal{C}$ de p.f;, l'un d'eux prolonge déjà $\mathcal{C}$. Si on représente ce dernier comme quotient de $\mathcal{O}^n$ par une limite de sous-faisceaux de type fini de $\mathcal{O}^n$, on voit de même qu'on peut remplacer $\mathcal{C}_1$ par un faisceau de p.f.

<u>Cor 1</u>. <u>Pour qu'un foncteur contravariant additif</u> F : (q coh) $\longrightarrow$ Ab <u>soit représentable</u>, <u>il faut et suffit qu'il soit èxact à gauche et transforme</u> $\varinjlim$ <u>filtrante en</u> $\varprojlim$ <u>filtrante</u>.

<u>Cor 2</u>. <u>Tout faisceau de p.f. défini sur un ouvert quasi-compact</u>

U de X se prolonge en un faisceau de p.f. sur X .

Soit $\mathcal{A}$ une catégorie abélienne. On aura à travailler de façon essentielle dans pro $D^b\mathcal{A}$, sous-catégorie pleine de pro $D(\mathcal{A})$ formée des "$\varprojlim$" K_i où la cohomologie de K est uniformément bornée avec i . H^p se prolonge en un foncteur pro $D(\mathcal{A}) \longrightarrow$ pro $\mathcal{A}$.

Proposition 3. Les foncteurs H^p forment un système conservatif (4) de foncteurs sur pro $D^b\mathcal{A}$.

Soit $f : K \longrightarrow L$ une flèche de pro $D^b\mathcal{A}$ telle que les $H^p(K) \longrightarrow H^p(L)$ soient des isomorphismes (dans pro $\mathcal{A}$). Il s'agit de vérifier que pour tout M dans $D^b(\mathcal{A})$, $Hom(K,M) \longleftarrow Hom(L,M)$ est un isomorphisme. Pósons K = "$\varprojlim$"K_α . On dispose d'une suite spectrale convergente

$Ext^p(H^qK_\alpha, M) \Longrightarrow Ext^{p+q}(K_\alpha, M)$ d'où par passage à la limite $\varinjlim Ext^p(H^qK_\alpha, M) \Longrightarrow \varinjlim Ext^{p+q}(K_\alpha, M)$, encore convergente car q est uniformément borné. Utilisant que tout foncteur (ici Ext) passe aux pro-objets et le foncteur $\varinjlim$: Ind Ab $\longrightarrow$ Ab , on peut réécrire :

$\varinjlim Ext^p(H^qK,M) \Longrightarrow \varinjlim Ext^{p+q}(K,M) = Hom(K[-p-q], M)$

Cette suite spectrale reste d'ailleurs vraie pour $M \in D(\mathcal{A})$. La proposition résulte aussitôt de l'existence de ces suites.

---

(4) i.e. si u est une flèche telle que les $H^p(u)$ soient inversibles, u est inversible.

n° 2. Lemmes fondamentaux.

Rappelons que les préschémas considérés sont noethériens.

<u>Proposition 4</u>. (Théorème de dualité pour une immersion ouverte).

<u>Soient</u> U <u>un ouvert de</u> X , j <u>la flèche d'inclusion</u>, $\mathcal{J}$ <u>un idéal définissant le fermé complémentaire</u>, $\mathcal{F}$ <u>un faisceau cohérent sur</u> U <u>et</u> $\mathcal{G}$ <u>quasi-cohérent sur</u> X . <u>Soit</u> $\overline{\mathcal{F}}$ <u>un prolongement cohérent de</u> $\mathcal{F}$. <u>On a</u>

$$\underline{\mathrm{Hom}_U(\mathcal{F}, j^*\mathcal{G}) = \mathrm{Hom}_X(\text{"}\varprojlim\text{"}\, \mathcal{J}^n\overline{\mathcal{F}}, \mathcal{G})} \qquad (5)$$

La flèche $\varinjlim \mathrm{Hom}_X(\mathcal{J}^n\overline{\mathcal{F}}, \mathcal{G}) \longrightarrow \mathrm{Hom}_U(\mathcal{F}, \mathcal{G})$ est injective : si l'image par f de $\mathcal{J}^n\overline{\mathcal{F}}$ dans $\mathcal{G}$ a son support dans le complémentaire de U , elle est annulée par une puissance de $\mathcal{J}$, soit $\mathcal{J}^k$, et l'image de $\mathcal{J}^{n+k}\overline{\mathcal{F}}$ est nulle.

Supposons maintenant X affine et soit $f \in \mathrm{Hom}_U(\mathcal{F}, \mathcal{G})$. Pour k assez grand, toute section s de $\mathcal{J}^k\overline{\mathcal{F}}$ sur X a une image (sur U) qui se prolonge en une section de $\mathcal{G}$ sur X . Remplaçant $\overline{\mathcal{F}}$ par $\mathcal{J}^k\overline{\mathcal{F}}$, on peut supposer disposer d'un diagramme

$$\begin{array}{ccccccccc} 0 & \longrightarrow & R & \longrightarrow & \mathcal{O}^n & \longrightarrow & \overline{\mathcal{F}} & \longrightarrow & 0 \\ & & \vdots\downarrow & & \| & & \vdots\downarrow & & \\ 0 & \longrightarrow & R_1 & \longrightarrow & \mathcal{O}^n & \longrightarrow & \mathcal{G} & & \end{array}$$

---

(5) Cette formule est évidemment bien connue, cf. p. ex. la thèse de P. GABRIEL.

où les flèches pointillées sont définies sur U .

Pour ℓ assez grand, en vertu de Krull et du fait que $R/R \cap R_1$ est concentré hors de U , on a $R \cap \mathcal{J}^{\ell}.\mathcal{O}^n \subset R_1$. , ce qui permet de prolonger f en $\bar{f} : \mathcal{J}^{\ell}\mathcal{F} \to \mathcal{G}$.

Dans le cas général U_i un recouvrement affine fini de X et U_{ijk} des recouvrements affins finis des $U_{ij} = U_i \cap U_j$. Les $\varinjlim$ commutent aux produits finis, d'où :

$$\begin{array}{ccccccc}
0 \to \mathrm{Hom}_U(\mathcal{F}, j^*\mathcal{G}) & \to & \prod_i \mathrm{Hom}_{U_i \cap U}(\mathcal{F}, j^*\mathcal{G}) & \rightrightarrows & \prod_{ijk} \mathrm{Hom}_{U_{ijk} \cap U}(\mathcal{F}, j^*\mathcal{G}) \\
\uparrow & & \uparrow \wr & & \uparrow \wr \\
0 \to \varinjlim \mathrm{Hom}_X(\mathcal{J}^n\mathcal{F}, j^*\mathcal{G}) & \to & \prod_i \varinjlim \mathrm{Hom}_{U_i}(\mathcal{J}^n\mathcal{F}, \mathcal{G}) & \rightrightarrows & \prod_{ijk} \varinjlim \mathrm{hom}_{U_{ijk}}(\mathcal{J}^n\mathcal{F}, \mathcal{G})
\end{array}$$

ce qui achève la démonstration.

La proposition 4 donne une formule explicite pour le foncteur (coh sur U) $\longrightarrow$ pro (coh sur X) "adjoint" à gauche à j^* . On notera ce foncteur $j_!$ ("prolongement par zéro") (6) ; il résulte de Krull qu'il est exact ; cela résulte aussi de ce que, X étant noethérien, un injectif de la catégorie des faisceaux quasi-cohérents sur X , restreint à U , reste injectif quasi-cohérent sur U .

<u>Proposition 5</u>. (Indépendance de la compactification).

$$\begin{array}{ccc}
U & \hookrightarrow & X \\
\| & & \downarrow f \\
U & \hookrightarrow & Y
\end{array}$$

<u>Soit</u> $f : X \longrightarrow Y$ <u>un morphisme propre, induisant un isomorphisme entre l'ouvert</u> U <u>de</u> Y

---

(6) Cette terminologie et notation conflictent évidemment avec ceux généralement admis (livre de Godement, SGA 1962 ...). Le lecteur préfèrera peut-être lire $\mathfrak{j}_!$ au lieu de $j_!$

et $f^{-1}(U)$. *Soit* $\mathcal{J}$ *un faisceau d'idéaux définissant* $Y - U$, *et soit* $\mathcal{J} = f^*\mathcal{J}$. *Si* $\mathcal{F}$ *est cohérent sur* X , *on a*

(i) *si* $k > 0$, *le système projectif* $R^kf_*\mathcal{J}^n\mathcal{F}$ *est essentiellement nul* (i.e. définit le proobjet nul)

(ii) *si* $k = 0$, *pour* n *assez grand*, $f_*\mathcal{J}^{n+1}\mathcal{F} = \mathcal{J}.f_*\mathcal{J}^n\mathcal{F}$

On a

$$\sum_{n=0}^{\infty} R^kf_*\mathcal{J}^n\mathcal{F} \text{ est un } \sum_{n=0}^{\infty}\mathcal{J}^n\text{-module de type fini (EGA III)}$$

donc pour n assez grand, $\mathcal{J}\otimes R^kf_*\mathcal{J}^n\mathcal{F} \longrightarrow R^kf_*\mathcal{J}^{n+1}\mathcal{F}$ est surjectif ; (ii) en résulte. Si $k>0$, R^kf_* est nul sur U ; il existe N tel que $\mathcal{J}^N.R^kf_*\mathcal{J}^n\mathcal{F}$ soit nul (quel que soit n).

Le diagramme

$$\begin{array}{ccccl} \mathcal{J}^p\otimes R^kf_*\mathcal{J}^n\mathcal{F} & \longrightarrow & R^kf_*\mathcal{J}^{n+p}\mathcal{F} & \longrightarrow 0 & (\text{si } n \; n_o) \\ \downarrow & & \downarrow & & \\ \mathcal{J}^p.\,R^kf_*\mathcal{J}^n\mathcal{F} & \longrightarrow & R^kf_*\mathcal{J}^n\mathcal{F} & & \end{array}$$

prouve que pour N assez grand et $k>0$,

$$R^kf_*\mathcal{J}^{n+N}\mathcal{F} \longrightarrow R^kf_*\mathcal{J}^n\mathcal{F} \text{ est nul, soit (i) .}$$

n° 3. Définition de $Rf_!$.

Un morphisme f (resp un couple de morphismes composables, resp. un triple ...) sera dit compactifiable si on peut respectivement trouver des diagrammes commutatifs

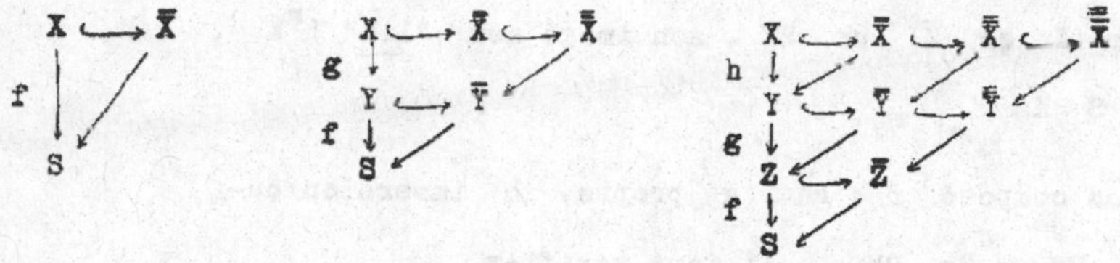

où les flèches horizontales sont des immersions ouvertes et les flèches obliques sont propres. Tout morphisme compactifiable est séparé de type fini, et Nagata affirme dans [2] que sa démonstration prouve la réciproque pour les schémas noethériens intègres, mais les hypothèses qu'il fait sur la base ne sont pas claires (7).

On se propose, pour tout f compactifiable, de définir $Rf_! : \mathrm{pro}D^b_{coh}(X) \longrightarrow \mathrm{pro}D^b_{coh}(S)$. Rappelons que la catégorie $D^b_{coh}(S)$, catégorie dérivée bornée de la catégorie des faisceaux cohérents sur S , est sous-catégorie pleine de la catégorie dérivée de celle de tous les faisceaux (non nécessairement quasi-cohérents) sur S , l'image essentielle étant formée des complexes à cohomologie cohérente et bornée.

Si f est propre, on dispose de $Rf_* : D^b_{coh}(X) \longrightarrow D^b_{coh}(S)$, de dimension finie, qui induit $Rf_! : \mathrm{pro}D^b_{coh}(X) \longrightarrow \mathrm{pro}D^b_{coh}(S)$. Si f est une immersion ouverte, $f_! :$ (coh sur X) $\rightarrow$ pro(coh sur S) induit D^b(coh sur X) $\rightarrow$ D^b pro (coh sur S) qui s'envoie dans pro $D^b_{coh}(S)$, et cette flèche se prolonge en $Rf_! : \mathrm{pro}D^b_{coh}(X) \longrightarrow \mathrm{pro}D^b_{coh}(S)$. Si on prolonge le complexe

---

(7) Mumford aurait vérifié via la démonstration de Nagata que tout morphisme séparé de type fini des schémas noethériens est compactifiable.

cohérent K sur X en $\bar{K}$ sur S , son image est $\text{"}\varprojlim\text{"}\, \mathcal{J}^n \bar{K}$, où $\mathcal{J}$ définit $S \smallsetminus X$.

Pour un composé $f = gh$ (g propre, h immersion ouverte) on prend $Rf_! = Rg_!\, Rh_!$. Il faut vérifier.

<u>Indépendance de la compactification</u>.

Considérons un diagramme commutatif

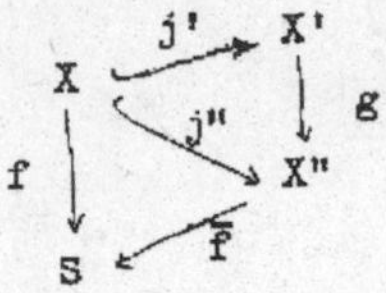

où j' et $j"$ sont des immersions ouvertes, $\bar{f}$ et g des applications propres. $j'X = g^{-1} j"X \cap j'X$. Il faut prouver que $R\bar{f}_* Rj"_! = R(\bar{f}g)_* Rj'_!$, soit encore, puisque $R(\bar{f}g)_* = R\bar{f}_* Rg_*$, que $Rj"_! = Rg_* Rj'_!$. On se ramène à supposer X dense dans X' , pour être dans les conditions de la proposition 5. Soit K un complexe cohérent borné sur X , prolongé en $\bar{K}$ sur X' . $g_*\bar{K}$ prolonge K sur $X"$, et on a une flèche

$$g_* \mathcal{J}'^n \bar{K} \longrightarrow Rg_* \mathcal{J}'^n \bar{K} \qquad (1)$$

où ' désigne un faisceau d'idéaux qui définit X' X . La prop. 5 (ii) montre que $j"_! K = \text{"}\varprojlim\text{"}\, g_* \mathcal{J}'^n \bar{K}$. Par passage à la "limite", on trouve $\text{"}\varprojlim\text{"}\, R^p g_* \mathcal{J}'^{n-q} \bar{K} \Longrightarrow \text{"}\varprojlim\text{"}\, R^{p+q} g_* \mathcal{J}'^n \bar{K}$, la proposition 5 (i) montre que cette suite spectrale dégénère, et il résulte de la prop. 3 que la flèche déduite de (1) par passage à la "limite" est un isomorphisme, d'où la formule voulue.

Il reste à vérifier

a) pour tout diagramme, on a une compatibilité

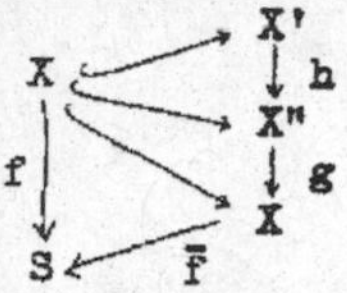

$$R\bar{f}_* Rj'''_! = R\bar{f}_* Rg_* Rj''_! = R\bar{f}_* Rg_* Rh_* Rj'_!$$

$$= R\bar{f}_* Rgh_* Rj'_! = Rf_* Rj'''_!$$

Comme deux compactifiactions peuvent toujours être coiffées par une troisième (par exemple $X' \underset{S}{x} X''$) , ceci suffit à prouver l'indépendance de l'arbitraire.

b) pour tout couple compactifiable, une identification $Rfg_! = Rf_! Rg_!$

c) pour tout triple compactifiable, une compatibilité

$$\begin{array}{ccc} Rfgh_! & = & Rfg_! \, Rh_! \\ \| & & \| \\ Rf_! Rgh_! & = & Rf_! \, Rg_! \, Rh_! \end{array}$$

On aurait alors prouvé

<u>Théorème 1</u>. <u>Pour</u> $f : X \longrightarrow Y$ <u>compactifiable</u>, $K \in \mathrm{proD}^b_{coh}(X)$, $L \in \mathrm{proD}^b_{coh}(Y)$ <u>posons</u> $\mathrm{Hom}_f(K,L) = \mathrm{Hom}_Y(Rf_! K, L)$. <u>Modulo des questions de compactificabilité</u>, <u>on fait ainsi des catégories</u> $\mathrm{proD}^b_{coh}(X)$ <u>une catégorie cofibrée sur les préschémas noethériens</u> (les flèches étant compactifiables ...)

n° 4. Définition de $Rf^!$.

Il résulte d'un tapis général de Verdier [1] que le foncteur $Rf_* : D(X) \longrightarrow D(S)$ a un adjoint à droite $D^+(S) \longrightarrow D^+(X)$; il ne mérite un nom, soit $Rf^!$, que pour f propre, et sa définition ne se prête pas directement au calcul.

Il faut tout d'abord expliciter un procédé de calcul de Rf_* au niveau des complexes, du type $Rf_*K = f_*C^*(K)$ où C^* est une résolution acyclique finie dépendant de façon fonctorielle, exacte, et compatible avec les limites inductives filtrantes de l'objet auquel on l'applique. A ce moment, pour tout injectif I sur S , $Hom(f_*C^p\mathcal{F},I)$ est exact en quasi-cohérent sur X , et transforme $\varinjlim$ en $\varprojlim$, donc est représentable par $f_p^! I$, injectif quasi-cohérent. Les $f_p^! I$ forment un complexe, ce qui définit $Rf^! : D^+S \longrightarrow D^+X$, adjoint à droite à $Rf_* : D(X) \longrightarrow D(S)$.

Voici comment définir une telle résolution :

1) Si S est séparé : on prend un recouvrement fini de X par des ouverts affines sur S (par exemple affines) et $Rf_*K = f_*\check{C}(\mathcal{U},K)$ (complexe de Čech alterné)

2) Sinon : les résolutions flasques canoniques ont ici deux défauts : a) $f_*C^*\mathcal{F}$ n'est plus quasi-cohérent ; cela ne porte pas à conséquence tant que, comme ici, les injectifs quasi-cohérents sont injectifs en tant que faisceaux.

b) C^* ne commute pas aux limites inductives filtrantes. Il est facile de corriger ce défaut : on représente quasi-cohérent comme $\mathcal{F} = \varinjlim \mathcal{F}_i$ ($\mathcal{F}_i$ de présentation finie) et on pose $C'^*\mathcal{F} = \varinjlim C^*\mathcal{F}_i$. Cela ne dépend pas de l'arbitraire.

Pour une immersion ouverte, on prendra $Rf^!K = f^*K$. Pour un morphisme composé $f = gh$ (g propre, h immersion ouverte), on prendra $Rf^! = Rg^! \, Rh^!$. Mettant bout à bout la proposition 4 et ce qui précède, on trouve

<u>Théorème 2</u>. (Dualité) <u>Soit</u> $f : X \longrightarrow S$ <u>compactifiable</u>, $f = gh$. <u>Les foncteurs</u> $Rf_! : \mathrm{pro}D^b_{coh}(X) \longrightarrow \mathrm{pro}D^b_{coh}(S)$ <u>et</u> $Rf^! : D^+(S) \longrightarrow D^+(X)$ <u>sont "adjoints" l'un de l'autre</u>.

Il faudrait vérifier l'indépendance de la compactification et un sorite de compatibilités et d'identifications analogue au cas de $Rf_!$. Des cas particuliers du formulaire résultent de la formule d'adjonction (unicité d'un vrai adjoint ...)

On pourra dans le cas général s'appuyer sur la proposition suivante :

<u>Proposition 6</u>. <u>Soit</u> $f : X \longrightarrow S$ <u>compactifiable et</u> $L \in D^+_{qc}(S)$. <u>La fibre en</u> $x \in X$ <u>de</u> $\mathcal{H}^p Rf^! L$ <u>est donnée par</u>

$$(\mathcal{H}^p Rf^! L)_x = \varinjlim_{x \in U} \mathrm{Hom}^p_S (Rf_!(U, \mathcal{O}_U), L)$$

On peut supposer X propre et L donné comme complexe de faisceaux injectifs quasi-cohérents. Soient U un ouvert affine de

X et $\mathcal{J}$ un idéal qui définit $X \setminus U$.

$$(\mathcal{H}^p Rf^! L)(U) = H^p(Rf^! L(U)) = H^p \, Hom_U (\mathcal{O}, Rf^! L)$$

$$= H^p \varinjlim Hom_X (\mathcal{J}^n, Rf^! L) = \varinjlim Hom^p_S (Rf_* \mathcal{J}^n, L)$$

$$= Hom^p_S (Rf_!(U, \mathcal{O}_U), L)$$ – La proposition en résulte.

On supposera admis pour $Rf^!$ un formalisme de variance analogue à celui du théorème 1 pour $Rf_!$.

Si on veut rendre plus explicite la définition de $Rf^!$ dans le cas propre, on peut remarquer que si $\mathcal{F}$ quasi-cohérent représente un foncteur F , pour tout ouvert U et $\mathcal{J}$ définissant $X \setminus U$, on a $\mathcal{F}(U) = \varinjlim Hom (\mathcal{J}^n, \mathcal{F}) = \varinjlim F (\mathcal{J}^n)$.

n° 5. Calcul de $Rf^!$.

Pour un morphisme fini, $Rf^!$ est ce qu'on veut ; pour le voir, il suffit de prendre pour foncteur résolvant, dans le calcul de Rf_* , l'identité. Pour l'espace projectif, l'unicité d'un adjoint ne laisse pas le choix (on utilise le fait qu'on a la formule d'adjonction pour $Rf_! = Rf_* : D(X) \longrightarrow D(S)$), on reviendra sur la flèche. Pour une immersion ouverte, $Rf^! = Rf^*$. Pour tout ouvert quasi-projectif U de X au-dessus de S , on peut donc calculer la restriction à U de $Rf^!$. En particulier, si on peut vérifier localement que $R^q f^! K = 0$ pour $q \neq p$, on connaît $Rf^! K$, puisqu'un objet de la catégorie dérivée n'ayant qu'un faisceau de cohomologie non nul est déterminé par ce faisceau. Rappelons un cas important où cette condition est vérifiée, avec $K = \mathcal{O}$.

Proposition 7. Soit $f : X \longrightarrow Y$, compactifiable. Supposons que localement (sur X et Y) on puisse trouver un diagramme

$$\begin{array}{ccc} X & \xrightarrow{f} & Y \\ & {}_{g}\searrow \quad \swarrow_{h} & \\ & S & \end{array}$$

où g et h sont localement intersection complète de dimension relative d' et d'' , et soit $d = d' - d''$.

Alors $Rf^{!}\mathcal{O}$ est réduit à un seul faisceau de cohomologie, situé en degré $-d$, qui est inversible.

En effet, $Rg^{!}\mathcal{O} = Rf^{!}\, Rh^{!}\mathcal{O}$, on sait que $Rg^{!}\mathcal{O}$ et $Rh^{!}\mathcal{O}$ sont du type indiqué, situés en degré $-d'$ et $-d''$, par des calculs locaux dans l'espace projectif; tout étant local, on peut supposer $R^{-d''}h^{!}\mathcal{O}$ isomorphe à $\mathcal{O}$, et l'assertion en résulte.

Pour passer de là à des complexes plus généraux, il faudrait au préalable démontrer des formules du type

$Rf^{!}(K \overset{L}{\otimes} L) = Rf^{!}(K) \overset{L}{\otimes} Lf^{*}L$. Seule la définition de la flèche pose un problème, l'isomorphie étant un problème local. Il faut bien sûr des restrictions de degré. Je n'ai rien vérifié en détail de ce qui suit.

1ère méthode pour définir la flèche.

Supposons K et L dans $D^{b}_{coh}(S)$ (et pas seulement quasi-cohérents), et en outre que

a) $Rf^!K$ est à degré borné (il est automatiquement cohérent, c'est un problème local)

b) $K \overset{L}{\otimes} L$ est à degré borné

c) $Rf^!K \overset{L}{\otimes} Lf^*L$ est à degré borné.

Pour définir $Rf^!K \overset{L}{\otimes} Lf^*L \longrightarrow Rf^!(K \overset{L}{\otimes} L)$, il suffit par dualité de définir $Rf_!(Rf^!K \overset{L}{\otimes} Lf\ L) \longrightarrow K \overset{L}{\otimes} L$, et par Kunneth $Rf_!(Rf^!K \overset{L}{\otimes} Lf^*L) = Rf_!\ Rf^!K \overset{L}{\otimes} L$; la flèche cherchée est induite par la flèche d'adjonction $Rf_!Rf^!K \longrightarrow K$.

<u>2ème méthode</u>.

On suppose f compactifiable, plat et localement d'intersection complète relative. On a alors $Rf^!\mathcal{O} = \omega[d]$, ω étant inversible. Désignons par $Rf_!$ un procédé de calcul de $Rf_!$ au niveau des complexes (c-à-d un tel procédé pour un compactifié $\bar{f} : \bar{X} \longrightarrow S$, de f), qu'on suppose du type décrit au n° 4, et tel que $C^p\mathcal{F}$ soit nul pour $p > d$ (ce qui peut s'obtenir par tronquage). On a alors une flèche $R^d\bar{f}_*\mathcal{F}[-d] \longrightarrow R\bar{f}_*\mathcal{F}$ pour tout $\mathcal{F}$ sur $\bar{X}$ (le dernier faisceau de cohomologie s'envoie dans son complexe). On en tire des morphismes de complexe
$\mathrm{Hom}_S\,(Rf_*\mathcal{F},I) \longrightarrow \mathrm{Hom}_S\,(R^d\bar{f}_*\mathcal{F}[-d],I)$ pour tout complexe I sur S . Soit I un injectif . $R^d\bar{f}_*\mathcal{F}[-d]$ est exact à droite en $\mathcal{F}$, le foncteur en $\mathcal{F}$ à droite de la flèche est donc représentable : on définit I_1 par

$$\mathrm{Hom}_S\,(R^df_*\mathcal{F},I) = \mathrm{Hom}_{\bar{X}}\,(\mathcal{F},I_1) \quad \text{et} \quad Rf_1^!I = I_1[d]$$

Ce qui précède définit

$$Rf^! I \longrightarrow Rf_1^! I$$

et pour un complexe d'injectifs, $Rf_1^!$ se définit terme à terme.

Il reste à évaluer $Rf_1^! I$ sur des ouverts affines, soit U, de X ; c'est donné par $\varinjlim \mathrm{Hom}\,(R^d f_! \mathcal{J}^n, I)$. Ce problème étant local en haut, il est facile de montrer par des méthodes projectives que le dual de "$\varprojlim$" $R^d f_! \mathcal{J}^n = R^d f_! (U, \mathcal{O})$, un ind-objet, est $f_* (U, \omega)$, représenté comme limite de ses sous-modules de sections ayant un pôle d'ordre ou plus n hors de $U (n \to \infty)$.

Ceci donne $Rf_1^! I = \omega \otimes f^* I[d]$, et la formule générale $Rf^! K = fK \otimes \omega[d]$.

BIBLIOGRAPHIE

[1] J.L. VERDIER, Séminaire Bourbaki, Novembre 65, exposé 300.

[2] M. NAGATA, Imbedding of an abstract variety in a complete variety. Journal of Math. of Kyoto University, vol. 2 n°1 1962.

ERRATA

Page 10, line 1: read "see Appendix" instead of "unpublished".

Page 85, Proposition 1.1: For the case of Qco(X), one needs the hypothesis "X quasi-separated".

Page 88, line 5: read "proper" instead of "of finite type".

Page 120, Ch II § 7: It has recently come to my attention that S. Kleiman has independently arrived at the results of [II 7.8] and [II 7.11] (unpublished).

Page 137, line 5: refer to [EGA IV § 16].

Page 144, line 4: read "invertible sheaf shifted n places to the left, where n = rel. dim X/Y" instead of "invertible sheaf".

Page 185, line 9: observe that the condition "f finite" is a consequence of the other conditions.

Page 190, line 7: read "a unique isomorphism" instead of " an isomorphim ".

Page 199: add at bottom: "Furthermore, the residue symbol is uniquely characterized by the properties (R0), (R1), (R2), (R5), (R6), and (R7)."

Page 249, line 8: "is" instead of "in".

Page 254, line 8: " $\mathbb{Z}$ instead of " $\mathbb{Q}$".

Page 288, Theorem 8.3: One must assume also that the fibres of f are of bounded dimension, so that $f^{\#}(R^{\cdot})$ will be in D^{+}.

Page 297, line 7: read "strengthens " instead of "generalizes".

Page 298, line 7: insert "shifted" after "invertible sheaf".

Page 301, line 5: "not" instead of "now".

Page 301, Problem 2: Giraud [6] provides us with an element of

$$H^2(X, \mathbb{G}_m)$$

(cohomology in the Zariski topology), whose vanishing is a necessary and sufficient condition for the global existence of a dualizing complex, supposed to exist locally. On the other hand, it is easy enough to construct a noetherian scheme of finite Krull dimension, which has a dualizing complex locally, but none globally, due to the lack of a global codimension function.

Page 373, line 1: change reference to [EGA V].

Page 403, last line: read "Publ. Math. I.H.E.S. no 29".

Printed in the United States
By Bookmasters